Health Implications of Perchlorate Ingestion

Committee to Assess the Health Implications
of Perchlorate Ingestion

Board on Environmental Studies and Toxicology

Division on Earth and Life Studies

NATIONAL RESEARCH COUNCIL
OF THE NATIONAL ACADEMIES

THE NATIONAL ACADEMIES PRESS
Washington, D.C.
www.nap.edu

THE NATIONAL ACADEMIES PRESS 500 Fifth Street, NW Washington, DC 20001

NOTICE: The project that is the subject of this report was approved by the Governing Board of the National Research Council, whose members are drawn from the councils of the National Academy of Sciences, the National Academy of Engineering, and the Institute of Medicine. The members of the committee responsible for the report were chosen for their special competences and with regard for appropriate balance.

This project was supported by Contract 68-C-03-081 between the National Academy of Sciences and the U.S. Environmental Protection Agency, U.S. Department of Defense, U.S. Department of Energy, and National Aeronautics and Space Administration. Any opinions, findings, conclusions, or recommendations expressed in this publication are those of the author(s) and do not necessarily reflect the view of the organizations or agencies that provided support for this project.

International Standard Book Number 0-309-09568-9 (Book)
International Standard Book Number 0-309-54811-X (PDF)

Library of Congress Control Number 2005925051

Additional copies of this report are available from

The National Academies Press
500 Fifth Street, NW
Box 285
Washington, DC 20055

800-624-6242
202-334-3313 (in the Washington metropolitan area)
http://www.nap.edu

Printed in the United States of America

THE NATIONAL ACADEMIES

Advisers to the Nation on Science, Engineering, and Medicine

The **National Academy of Sciences** is a private, nonprofit, self-perpetuating society of distinguished scholars engaged in scientific and engineering research, dedicated to the furtherance of science and technology and to their use for the general welfare. Upon the authority of the charter granted to it by the Congress in 1863, the Academy has a mandate that requires it to advise the federal government on scientific and technical matters. Dr. Bruce M. Alberts is president of the National Academy of Sciences.

The **National Academy of Engineering** was established in 1964, under the charter of the National Academy of Sciences, as a parallel organization of outstanding engineers. It is autonomous in its administration and in the selection of its members, sharing with the National Academy of Sciences the responsibility for advising the federal government. The National Academy of Engineering also sponsors engineering programs aimed at meeting national needs, encourages education and research, and recognizes the superior achievements of engineers. Dr. Wm. A. Wulf is president of the National Academy of Engineering.

The **Institute of Medicine** was established in 1970 by the National Academy of Sciences to secure the services of eminent members of appropriate professions in the examination of policy matters pertaining to the health of the public. The Institute acts under the responsibility given to the National Academy of Sciences by its congressional charter to be an adviser to the federal government and, upon its own initiative, to identify issues of medical care, research, and education. Dr. Harvey V. Fineberg is president of the Institute of Medicine.

The **National Research Council** was organized by the National Academy of Sciences in 1916 to associate the broad community of science and technology with the Academy's purposes of furthering knowledge and advising the federal government. Functioning in accordance with general policies determined by the Academy, the Council has become the principal operating agency of both the National Academy of Sciences and the National Academy of Engineering in providing services to the government, the public, and the scientific and engineering communities. The Council is administered jointly by both Academies and the Institute of Medicine. Dr. Bruce M. Alberts and Dr. Wm. A. Wulf are chair and vice chair, respectively, of the National Research Council

www.national-academies.org

Committee to Assess the Health Implications of Perchlorate Ingestion

Members

Richard B. Johnston, Jr. (*Chair*), University of Colorado School of Medicine, Denver
Stacy Branch, Djehuty Biomed Consulting, Clayton, NC
Gregory Brent, University of California, Los Angeles
Rosalind Brown, Harvard University School of Medicine and Children's Hospital in Boston, MA
Charles C. Capen, The Ohio State University, Columbus
David Cooper, The Johns Hopkins University; Sinai Hospital of Baltimore, MD
Richard Corley, Pacific Northwest National Laboratory, Richland, WA
Linda Cowan, University of Oklahoma, Oklahoma City
James C. Lamb, The Weinberg Group Inc., Washington, DC
George Lambert, UMDNJ–Robert Wood Johnson Medical School, Piscataway, NJ
R. Michael McClain, McClain Associates, Randolph, NJ
Susan Schantz, University of Illinois at Urbana-Champaign, Urbana
Dalene Stangl, Duke University, Durham, NC
Lynette Stokes, Environmental Health Administration, Department of Health, Washington, DC
Robert Utiger, Harvard University School of Medicine, Boston, MA

Staff

Ellen Mantus, Project Director
Roberta Wedge, Program Director for Risk Assessment
Mary Fox, Program Officer
Norman Grossblatt, Senior Editor
Jennifer Saunders, Research Associate
Robert Policelli, Project Assistant
Laura Waters, Project Assistant

Sponsors

U.S. Environmental Protection Agency
U.S. Department of Defense
U.S. Department of Energy
National Aeronautics and Space Administration

Board on Environmental Studies and Toxicology[1]

---

[1]This study was planned, overseen, and supported by the Board on Environmental Studies and Toxicology.

OTHER REPORTS OF THE BOARD ON ENVIRONMENTAL STUDIES AND TOXICOLOGY

Air Quality Management in the United States (2004)
Endangered and Threatened Species of the Platte River (2004)
Atlantic Salmon in Maine (2004)
Endangered and Threatened Fishes in the Klamath River Basin (2004)
Cumulative Environmental Effects of Alaska North Slope Oil and Gas Development (2003)
Estimating the Public Health Benefits of Proposed Air Pollution Regulations (2002)
Biosolids Applied to Land: Advancing Standards and Practices (2002)
The Airliner Cabin Environment and Health of Passengers and Crew (2002)
Arsenic in Drinking Water: 2001 Update (2001)
Evaluating Vehicle Emissions Inspection and Maintenance Programs (2001)
Compensating for Wetland Losses Under the Clean Water Act (2001)
A Risk-Management Strategy for PCB-Contaminated Sediments (2001)
Acute Exposure Guideline Levels for Selected Airborne Chemicals (4 volumes, 2000-2004)
Toxicological Effects of Methylmercury (2000)
Strengthening Science at the U.S. Environmental Protection Agency (2000)
Scientific Frontiers in Developmental Toxicology and Risk Assessment (2000)
Ecological Indicators for the Nation (2000)
Waste Incineration and Public Health (1999)
Hormonally Active Agents in the Environment (1999)
Research Priorities for Airborne Particulate Matter (4 volumes, 1998-2004)
The National Research Council's Committee on Toxicology: The First 50 Years (1997)
Carcinogens and Anticarcinogens in the Human Diet (1996)
Upstream: Salmon and Society in the Pacific Northwest (1996)
Science and the Endangered Species Act (1995)
Wetlands: Characteristics and Boundaries (1995)
Biologic Markers (5 volumes, 1989-1995)
Review of EPA's Environmental Monitoring and Assessment Program (3 volumes, 1994-1995)
Science and Judgment in Risk Assessment (1994)
Pesticides in the Diets of Infants and Children (1993)
Dolphins and the Tuna Industry (1992)
Science and the National Parks (1992)
Human Exposure Assessment for Airborne Pollutants (1991)
Rethinking the Ozone Problem in Urban and Regional Air Pollution (1991)
Decline of the Sea Turtles (1990)

Copies of these reports may be ordered from the National Academies Press
(800) 624-6242 or (202) 334-3313
www.nap.edu

Preface

In 1985, perchlorate contamination was discovered at Superfund sites in California; however, the extent of perchlorate contamination of water sources nationwide was not revealed until 1997. Today, over 11 million people have perchlorate in their public drinking-water supplies at concentrations of 4 ppb (4 μg/L) or higher. Because of the controversy surrounding the concentration at which perchlorate should be regulated, the Department of Defense (DOD), the Department of Energy (DOE), the National Aeronautics and Space Administration (NASA), and the Environmental Protection Agency (EPA) asked the National Research Council to assess the potential adverse health effects of perchlorate ingestion from clinical, toxicologic, medical, and public-health perspectives.

In this report, the Committee to Assess the Health Implications of Perchlorate Ingestion reviews the current state of the science regarding potential adverse health effects of perchlorate exposure. Specifically, the committee evaluated human clinical and epidemiologic studies and animal toxicology studies, and determined the relevance of the animal studies for predicting adverse effects in humans, especially sensitive populations. The committee also assessed perchlorate concentrations at which chronic inhibition of iodide uptake and subsequent changes in thyroid hormone production might lead to adverse health effects in humans. As a final task, the committee reviewed and determined whether EPA's findings in its 2002 draft risk assessment, *Perchlorate Environmental Contamination: Toxicological Review and Risk Characterization,* are consistent with current scientific evidence. Recommendations are provided for scientific research that could reduce uncertainty in the understanding of human health effects associated with low-level perchlorate ingestion.

This report has been reviewed in draft form by persons chosen for their diverse perspectives and technical expertise in accordance with procedures approved by the National Research Council's Report Review Committee. The purpose of this independent review is to provide candid and critical comments that will assist the perchlorate committee and the Research Council in making the published report as sound as possible and to ensure that the report meets the Research Council's standards of objectivity, evidence, and responsiveness to the study charge. The review comments and draft manuscript remain confidential to protect the integrity of the deliberative process. We wish to thank the following people for their review of this report: Michael Aschner, Vanderbilt University Medical Center; Gerard Burrow, Yale University School of Medicine; George Daston, Proctor and Gamble Company; Kelly Dix, Lovelace Respiratory Research Institute; Ronald Estabrook, The University of Texas Southwestern Medical Center; Ellen Gold, University of California, Davis; Philip Landrigan, Mount Sinai School of Medicine; Gilbert Omenn, University of Michigan Medical School; Louise Ryan, Harvard School of Public Health; Rudi Schmid, Retired; Jerrold Ward, National Institutes of Health; and E. Dillwyn Williams, University of Cambridge.

Although the reviewers listed above have provided many constructive comments and suggestions, they were not asked to endorse the conclusions or recommendations, nor did they see the final draft of the report before its release. The review of this report was overseen by John C. Bailar and Floyd Bloom. Appointed by the National Research Council, they were responsible for making certain that an independent examination of the report was carried out in accordance with institutional procedures and that all review comments were carefully considered. Responsibility for the final content of the report rests entirely with the committee and the National Research Council.

The committee gratefully acknowledges the persons who made presentations at the committee's public meetings (see Appendix C) and Raymond York, of Argus Laboratories, for providing the committee with original data from selected animal toxicology studies. The committee also thanks the sponsor representatives who responded to data requests and provided background materials: Lisa Matthews, Annie Jarabek, and William Farland, EPA; Daniel Rogers and Jeff Cornell, DOD; Richard Williams and Richard Wickman, NASA; and Patrice Bubar, Karen Guevara, Blaine Rowley, and Mark Frei, DOE.

The committee is especially grateful for the consistently strong and knowledgeable assistance of the National Research Council staff in preparing this report, particularly Ellen Mantus, project director, but also James Reisa, director of the Board on Environmental Studies and Toxicology; Roberta Wedge, program director for risk analysis; Mary Fox, program officer, Jennifer Saunders, research associate, Mirsada Karalic-Loncarevic, research associate; Ruth E. Crossgrove, senior editor; Norman Grossblatt, senior editor; Bryan Shipley, research associate; and Robert Policelli and Laura Waters, project assistants.

Finally, I thank the members of the committee for their commitment to the breadth and importance of our task and their dedication to deriving conclusions and recommendations based only on the best available scientific evidence.

Richard B. Johnston, Jr.
Chair, Committee to Assess the Health
Implications of Perchlorate Ingestion

Contents

FIGURES

TABLES

Health Implications of Perchlorate Ingestion

Summary

IN 1985, perchlorate contamination was discovered in wells at California Superfund sites; however, perchlorate contamination of water sources nationwide was not recognized until 1997. Today, more than 11 million people have perchlorate in their public drinking-water supplies at concentrations of at least 4 ppb (4 μg/L).[1] No national drinking-water standard for perchlorate exists, and the concentration at which a standard should be set is being debated. The U.S. Environmental Protection Agency (EPA), which has the responsibility for establishing national drinking-water standards, has issued draft risk assessments of perchlorate. However, the assessments have come under criticism on the grounds that the conclusions presented in them are based on flawed scientific studies and that not all available data have been incorporated appropriately into them.

In view of the controversy surrounding the concentration at which perchlorate should be regulated, EPA, the Department of Defense (DOD), the Department of Energy (DOE), and the National Aeronautics and Space Administration (NASA) asked the National Research Council (NRC) to assess independently the adverse health effects of perchlorate ingestion from clinical, toxicologic, and public-health perspectives. They also asked the NRC to evaluate the relevant scientific literature and key findings underlying EPA's 2002 draft risk assessment, *Perchlorate Environmental Contamination: Toxicological Review and Risk Characterization.*

---

[1]The estimate of 11 million people is based on sampling data collected as of May 2004 by the U.S. Environmental Protection Agency (EPA) as required by the Unregulated Contaminant Monitoring Rule. The minimum reporting level for that data collection is 4 parts per billion (ppb) (4 micrograms per liter [μg/L]).

THE CHARGE TO THE COMMITTEE

In response to the agencies' request, the NRC convened the Committee to Assess the Health Implications of Perchlorate Ingestion, which prepared this report. The members of the committee were selected for their expertise in pediatrics; endocrinology; pediatric endocrinology; thyroid endocrinology, physiology, and carcinogenesis; immunology; veterinary pathology; animal toxicology; neurotoxicology; developmental toxicology; physiologically based pharmacokinetic modeling; epidemiology; biostatistics; and risk assessment.

The committee was asked to assess the current state of the science regarding potential adverse effects of disruption of thyroid function by perchlorate in humans and laboratory animals at various stages of life. It was asked to evaluate the animal studies with particular attention to key end points, including changes in brain morphometry, behavior, thyroid hormone levels, and thyroid histopathology. On the basis of its review, the committee was asked to determine whether EPA's findings in its 2002 draft risk assessment, *Perchlorate Environmental Contamination: Toxicological Review and Risk Characterization*, are consistent with the current scientific evidence. The committee was also asked to suggest specific scientific research that could reduce the uncertainty in the understanding of human health effects associated with ingestion of low concentrations of perchlorate. The committee's complete statement of task is provided in Chapter 1 of this report.

THE COMMITTEE'S APPROACH TO ITS CHARGE

The committee held five meetings from October 2003 to July 2004. During public sessions at the first, second, and fourth meetings, the committee heard presentations from representatives of the U.S. Office of Science and Technology Policy, EPA, DOD, DOE, NASA, the Food and Drug Administration, the Agency for Toxic Substances and Disease Registry, the California Environmental Protection Agency, Congress, and other interested parties, including industry and environmental groups. At the second meeting, several noted scientists were invited to make presentations to the committee to answer questions raised at the first public meeting. The committee reviewed (1) materials submitted by EPA, DOD, DOE, NASA, industry, and private individuals, (2) studies evaluated in EPA's 2002 draft perchlorate risk assessment, (3) findings in EPA's 2002 draft perchlorate risk assessment, and (4) information from publicly available scientific literature. Accordingly, the committee evaluated both published and unpublished data; however, it typically gave more weight in its deliberations to published reports. Unpublished data were considered only when the committee had sufficient information to evaluate the methods used to produce them. Overall, emphasis was given to studies with the soundest scientific methods to draw conclusions regarding the effects of perchlorate exposure.

COMMITTEE'S EVALUATION AND FINDINGS

Thyroid Function

The thyroid gland produces two hormones, thyroxine (T_4) and triiodothyronine (T_3), which circulate in the bloodstream primarily bound to protein. Thyroid hormone synthesis and secretion are normally maintained within narrow limits by an efficient regulatory mechanism. Specifically, decreases in serum thyroid hormone concentrations lead to increases in the secretion of thyrotropin (thyroid-stimulating hormone, TSH) by the pituitary gland, and increases in serum thyroid hormone concentrations lead to decreases in TSH secretion. TSH stimulates virtually every step of thyroid hormone production and secretion, and such tight control of TSH secretion maintains thyroid hormone production and secretion within normal or nearly normal limits and thereby protects against both hypothyroidism (deficiency of thyroid hormone production) and hyperthyroidism (excess of thyroid hormone production).

T_4 is largely a precursor hormone with little or no intrinsic biologic activity, and it is converted to T_3, the biologically active thyroid hormone, in most tissues of the body. T_3 is required for normal development of the central nervous system in fetuses and infants. Its actions include stimulation of the development and growth of neurons (nerve cells) and glial (supporting) cells, the formation of synapses (connections) between neurons, the formation of the myelin sheaths that surround neuronal processes, and the development of neurotransmitters, which transmit signals from one neuron to another. T_3 is also required for normal skeletal development and growth. In both infants and adults, T_3 and T_4 are critical determinants of metabolic activity and affect the function of virtually every organ system.

Iodine is a component of T_4 and T_3, and transfer of iodide from the circulation into the thyroid gland is an essential step in the synthesis of the two hormones.[2] Iodide transport into the thyroid is mediated by a protein molecule known as the sodium (Na^+)/iodide (I^-) symporter (NIS). NIS molecules bind iodide with very high affinity, but they also bind other ions that have a similar shape and electric charge. The binding of those other ions to the NIS inhibits iodide transport into the thyroid, which can result in intrathyroid iodide deficiency and consequently decreased synthesis of T_4 and T_3. For adverse health effects to occur in healthy adults, thyroid hormone production must fall substantially and, more importantly, must remain low for at least several weeks. The minimal prolonged decrease in thyroid hormone production that would be associated with adverse health effects is not known; any decrease is potentially more likely to have adverse effects in sensitive populations (people with thyroid disorders, pregnant women, fetuses, and infants), but data are not available to determine the magnitude of the decrease needed to cause adverse effects in those populations.

Iodide can be obtained only by ingestion of food or water that contains it. Therefore, iodide deficiency and reduction in thyroid hormone production can occur if iodide intake is very low. Because the body maintains the serum concentrations of thyroid hormones within narrow limits through feedback control mechanisms, there is remarkable compensation for iodide deficiency. Generally, thyroid hormone production is normal even when iodide intake is quite low. Hypothyroidism occurs only if daily iodide intake is below about 10 to 20 μg (about one-fifth to one-tenth of the average intake in the United States). However, in pregnant women, iodide deficiency of that severity can result in major neurodevelopmental deficits and goiter in their offspring. Lesser degrees of iodide deficiency may also cause important neurodevelopmental deficits in infants and children.

Perchlorate and the Thyroid

Perchlorate can affect thyroid function because it is an ion that competitively inhibits the transport of iodide into the thyroid by the NIS. In the 1950s and 1960s, potassium perchlorate was given for long periods to patients who had hyperthyroidism to reduce T_4 and T_3 production by inhibiting thyroid iodide uptake. The medical literature of that era contains reports of successful treatment of more than 1,000 hyperthyroid patients

---

[2]Iodide is the negatively charged ion of iodine and is the form of iodine that is found in foods and in the circulation in humans.

with potassium perchlorate at doses of 400-2,000 milligrams (mg) per day for many weeks or months. Among them were 12 women who had hyperthyroidism and were treated with 600-1,000 mg of potassium perchlorate per day during pregnancy. One infant had slight thyroid enlargement that decreased soon after birth. No other abnormalities were reported; however, no thyroid-function tests or neurodevelopmental evaluations were conducted, and the infants were not followed thereafter.

Treatment of hyperthyroid patients with potassium perchlorate typically caused few side effects, although some patients had nausea, vomiting, rashes, fever, lymph node enlargement, or kidney dysfunction. The frequency of side effects was dose-dependent. Thirteen patients who had taken 400-1,000 mg of potassium perchlorate per day for 2-20 weeks developed aplastic anemia (cessation of production of red blood cells) or agranulocytosis (cessation of production of white blood cells), and seven of them died. Because of those events and the development of better antithyroid drugs, the use of perchlorate to treat hyperthyroid patients largely ceased by the late 1960s.

A study of long-term administration of potassium perchlorate reported in 1984, however, provides useful data. Eighteen people who had hyperthyroidism caused by Graves disease were treated initially with 900 mg per day. The dose of potassium perchlorate was reduced over a 12-month period to an average of 93 mg per day as thyroid function returned to normal. The patients then received 40-120 mg per day for 12 months. During that period, all the patients had normal serum T_4 and T_3 concentrations, and most patients no longer had high serum concentrations of TSH-receptor stimulating antibodies, which are the cause of hyperthyroidism in patients who have Graves disease. Absence of the antibodies indicated that the patients no longer had Graves disease. Thus, one could consider treatment in the latter 12 months to be equivalent to administration of perchlorate to healthy people. Therefore, the results provide evidence that moderately high doses of perchlorate given chronically to people with a history of hyperthyroidism do not cause hypothyroidism.

Overall, there have been no reports of the appearance of new thyroid disorders, thyroid nodules, or thyroid carcinomas in patients treated with potassium perchlorate for hyperthyroidism.

Perchlorate has been given to healthy men and women for up to 6 months to determine its effects on normal thyroid function. It is usually administered as potassium perchlorate, which is rapidly absorbed after ingestion and rapidly eliminated from the body unchanged primarily in urine. In those studies, perchlorate doses ranged from 0.007 to 9.2 mg per kilogram (kg) per day, assuming a 70-kg body weight. There were no changes in serum T_4, T_3, or TSH concentrations to suggest that thyroid function was adversely affected. The highest dose (9.2 mg/kg per day) lowered thyroid iodide content by 25% and resulted in a very small decrease in serum free T_4 concentrations after 4 weeks of perchlorate administration. However, serum TSH concentrations were also lower, not higher, as would occur if serum free T_4 concentrations decreased to an important extent. Some doses of perchlorate did inhibit thyroid uptake of radioiodide. For example, a 2-week study found that inhibition of radioiodide uptake ranged from 1.8% at 0.007 mg/kg per day to 67.1% at 0.5 mg/kg per day. Uptake at the lowest dose of 0.007 mg/kg per day was not significantly different from baseline in that study. In a 6-month study, there was also no inhibition in healthy subjects given 0.007 mg/kg per day.

Because of the body's compensation mechanisms, it is not likely that the decreases in thyroid iodide uptake reported in short-term studies would be sustained; rather, iodide uptake would be expected to return to normal. To cause declines in thyroid hormone production that would have adverse health effects, iodide uptake would most likely have to be reduced by at least 75% for months or longer. On the basis of the studies of long-term treatment of hyperthyroidism in which patients continued to be given perchlorate after their hyperthyroidism resolved and the clinical studies of healthy adults, the perchlorate dose required to cause hypothyroidism in adults would probably be more than 0.40 mg/kg per day, assuming a 70-kg body weight. However, in pregnant women, infants, children, and people with low iodide intake or pre-existing thyroid dysfunction, the dose required to cause hypothyroidism may be lower.

Epidemiologic Studies

Epidemiologic studies have examined the association of perchlorate exposure with thyroid function and thyroid disease in newborns, children, and adults. Only one study has examined the relationship between perchlorate exposure and adverse neurodevelopmental outcomes in children. Almost all the epidemiologic studies are "ecologic." In ecologic studies, exposure data, outcome data, or both are available only for large geographic areas, not for individuals. Exposures measured within areas are then applied to all persons living in the areas. The assumption that the geographically defined exposures reasonably represent those of all people in an area is more reasonable when one is studying agents that are parts of a common-source exposure, such as contaminants in a municipal water supply; however, individual water consumption is still likely to vary because of the use of well water or bottled water or a nonuniform distribution of contaminants within a geographic area. Ecologic studies do not include information about exposure *and* outcome *within* individuals, so they are considered to be the weakest type of observational studies. They are subject to what is referred to as the ecologic fallacy in that relationships observed (or not observed) between exposure and outcomes at the ecologic level may not apply at the individual level. Thus, ecologic studies alone cannot provide direct evidence of causation, although their results can provide supporting data concerning a possible causal relationship.

Acknowledging that ecologic data alone are not sufficient to demonstrate whether or not an association is causal, the committee found that they can provide evidence bearing on possible associations and reached the following conclusions regarding the proposed association of perchlorate exposure with various health end points:

- *Congenital hypothyroidism.* The available epidemiologic evidence is not consistent with a causal association between perchlorate exposure and congenital hypothyroidism as defined by the authors of the studies reviewed by the committee. All studies of that association were negative.
- *Changes in thyroid function in newborns.* The available epidemiologic evidence is not consistent with a causal association between exposure during gestation to perchlorate in the drinking water at up to 120 ppb and changes in thyroid hormone and TSH production in normal-birthweight, full-term newborns. Most of the studies show neither significantly lower T_4 production nor significantly higher TSH secretion in infants born in geographic areas in which the water supply had measurable perchlorate concentrations. However, no data are available on the association of perchlorate exposure with thyroid dysfunction in the groups of greatest concern, low-birthweight or preterm newborns, offspring of mothers who had iodide deficiency during gestation, or offspring of hypothyroid mothers.
- *Neurodevelopmental outcomes.* The epidemiologic evidence is inadequate to determine whether or not there is a causal association between perchlorate exposure and adverse neurodevelopmental outcomes in children. Only one pertinent study has been conducted: an ecologic study that examined the association of perchlorate exposure with autism and attention-deficit-hyperactivity disorder (ADHD). Although the committee considers the inclusion of ADHD plausible, it questions the appropriateness of autism as an end point given that autism has not been observed in the spectrum of

clinical outcomes in children who had congenital hypothyroidism and were evaluated prospectively.

- *Hypothyroidism and other thyroid disorders in adults.* The evidence from chronic, occupational-exposure studies and ecologic investigations in adults is not consistent with a causal association between perchlorate exposure at the doses investigated and hypothyroidism or other thyroid disorders in adults. In occupational studies, perchlorate doses as high as 0.5 mg/kg per day have not been associated with adverse effects on thyroid function in workers. However, the small sample sizes in some studies may have reduced the ability to identify important differences, and the studies were limited to those workers who remained in the workforce.
- *Thyroid cancer in adults.* The epidemiologic evidence is insufficient to determine whether or not there is a causal association between exposure to perchlorate and thyroid cancer. Only two pertinent ecologic studies have been conducted. In one, the number of cancer cases was too small to have a reasonable chance of detecting an association if one existed. In the second, people were exposed to both perchlorate and trichloroethylene. In neither study was it possible to adjust for potential confounding variables. However, the committee questions the biologic plausibility of thyroid cancer as a likely outcome of perchlorate exposure.

The committee emphasizes that no studies have investigated the relationship between perchlorate exposure and adverse outcomes among especially vulnerable groups, such as low-birthweight or preterm infants. The available studies did not assess the possibility of adverse outcomes in the offspring of mothers who were exposed to perchlorate and had a low dietary iodide intake. Finally, there have been no adequate studies of maternal perchlorate exposure and neurodevelopmental outcomes in infants.

Animal Toxicology Studies

The pituitary-thyroid system of rats is similar to that of humans. For example, decreases in thyroid hormone production result in increased secretion of TSH, which then increases thyroid production and release of T_4 and T_3. However, differences in binding proteins, binding affinities of the proteins for the hormones, turnover rates of the hormones, and thyroid stimulation by placental hormones lead to important quantitative differences between the two species. Therefore, although studies in rats provide useful qualitative information on potential adverse effects of perchlorate exposure, they are limited in their utility for quantitatively assessing human health risk associated with perchlorate exposure.

One of the most controversial issues regarding the animal toxicology studies is the interpretation of results of rat studies that evaluated the effects of maternal perchlorate exposure on offspring brain development. In those studies, female rats were given ammonium perchlorate throughout pregnancy and into the postnatal period. Linear measurements of several brain regions of the male and female pups at several postnatal ages were compared with control values. The most consistent change observed was a statistically significant increase in the width of the corpus callosum; however, the dose at which that change was observed was not consistent between studies. Serious questions have been raised regarding the design and methods used in those studies. The committee agrees with previous reviewers that the methodologic problems, such as possible systematic differences in the plane of section across treatment groups, and the lack of a consistent dose-response relationship make it impossible to conclude whether or not perchlorate exposure causes changes in brain structure. Furthermore, the committee notes two issues concerning study design. First, although it may be appropriate to collect data in a nonblinded fashion in a comparative-morphology study, it is *not* appropriate to collect measurements of linear thickness of tissues in a nonblinded fashion, as was done for a set of sections measured in those studies. Second, measurements of the thickness of brain areas are not the most sensitive method of detecting alterations in neural structure. Measurements of area or volume would be more sensitive and more accurate, particularly for structures, such as the corpus callosum, that change shape across serial coronal sections. Furthermore, all measurements of size, including measurements of area and volume, are only a surrogate for changes in the cellular structure of a brain region, and it is ultimately the underlying cellular changes and their neurodevelopmental effects that are important to understand.

Other studies that have received critical attention are rat studies that investigated the effect of maternal exposure on offspring neurobehavior. In the primary study, female rats were treated with ammonium perchlorate throughout pregnancy and into the postnatal period, and the offspring evaluated with a battery of behavioral tests. Overall, the committee found that the functions evaluated (motor activity, auditory startle, learning, and memory) were appropriate in light of the suspected mode of action of perchlorate. However, the tests used were screening measures and were unlikely to detect subtle alterations in motor or cognitive functions associated with moderate reductions in circulating thyroid hormones. In addition, some important functional end points—such as auditory function, balance, and coordination—were not assessed. Therefore, it is not surprising that no significant effects of perchlorate were observed on any of the behavioral measures except an increase in motor activity in male pups on one day of testing. Because the tests lacked the sensitivity to detect subtle effects, the committee concludes that the data are inadequate to determine whether or not gestational or lactational exposure to perchlorate affects behavioral function in rats.

Concerns have also been raised over the significance of the results of a two-generation rat study in which benign thyroid tumors were observed in two male offspring. Both the parent generation and the offspring were given ammonium perchlorate before mating, during mating, gestation, and lactation, and until sacrifice. The offspring had additional exposure during their gestation and lactation periods. On review of the original pathology data, the committee found that one control male rat in the parent generation also had a benign thyroid tumor at about the same age of those observed in the two male offspring. The committee agreed that the two male offspring did have benign thyroid tumors and noted that the observations are expected in rats given high concentrations of goitrogenic chemicals known to interfere with thyroid hormone homeostasis. The committee concludes that the thyroid tumors in the offspring were most likely treatment-related but that thyroid cancer in humans resulting from perchlorate exposure is unlikely because of the hormonally mediated mode of action and species differences in thyroid function.

On the basis of observations that high doses of perchlorate in humans with hyperthyroidism have caused side effects that could be considered immunologic responses—such as skin rashes, aplastic anemia, or agranulocytosis—some have suggested that perchlorate exposure might adversely affect the immune system. However, extensive immunotoxicity studies in mice revealed no changes in immunologic function in response to perchlorate exposure. Therefore, the committee finds that there is no evidence for a causative relationship between perchlorate ingestion and any biologically meaningful stimulatory or inhibitory effect on the immune system in rodents, and concludes that the side effects in humans were probably toxic effects of the very high doses of perchlorate given to those patients.

EPA's 2002 Perchlorate Risk Assessment

The committee's review of EPA's findings in its 2002 perchlorate risk assessment focused on four points: the mode-of-action model of perchlorate toxicity, the definition of adverse effect, the point of departure, and the use of uncertainty factors to derive a reference dose (RfD) for daily oral exposures to perchlorate. EPA's proposed mode-of-action model of perchlorate toxicity represents a continuum of possible health effects of perchlorate exposure. Ultimately, EPA's model shows birth defects in children and tumors in adults as possible effects of inhibition of thyroid iodide uptake. The committee finds that EPA's mode-of-action model adequately represents the possible early sequence of events after perchlorate exposure, but it does not think that the model is an accurate representation of possible outcomes after changes in thyroid hormone and TSH production. Figure S-1 shows the committee's suggested mode-of-action model of perchlorate toxicity in humans.

The committee concludes that the most reasonable pathway of events after sustained changes in thyroid hormone and TSH secretion would be thyroid hypertrophy or hyperplasia, possibly followed by hypothyroidism in people unable to compensate with an increase in thyroid iodide uptake. At that point, the pathway would diverge to two potential outcomes—metabolic sequelae (such as decreased metabolic rate and slowing of the function of many organ systems) at any age and abnormal growth and development of fetuses and children. The committee concludes that the development of thyroid tumors, as an ultimate result of perchlorate-caused inhibition of thyroid iodide uptake, is unlikely in humans.

An important point is that inhibition of thyroid iodide uptake is the only effect that has been consistently documented in humans exposed to perchlorate. Furthermore, the outcomes at the end of the continuum are not inevitable consequences of perchlorate exposure. Mechanisms exist that allow the body to compensate for decreases in T_4 and T_3 production. The compensatory increase in TSH secretion and thyroid iodide uptake can return T_4 and T_3 production to normal without causing adverse effects on human health.

Given the mode-of-action model shown in Figure S-1, the committee concludes that the first adverse effect in the continuum would be hypothyroidism. Any effects that follow and result from hypothyroidism clearly would be adverse. EPA defined changes in serum thyroid hormone and TSH concentrations as adverse effects. The committee does not agree that transient changes in serum thyroid hormone or TSH concentrations are adverse health effects; they are simply biochemical changes that might precede adverse effects.

A primary purpose of EPA's perchlorate risk assessment was to calculate an RfD. The first step in deriving an RfD is a comprehensive review of all relevant human and animal data. Traditionally, a critical effect and a critical study are then identified that serve as the point of departure for the

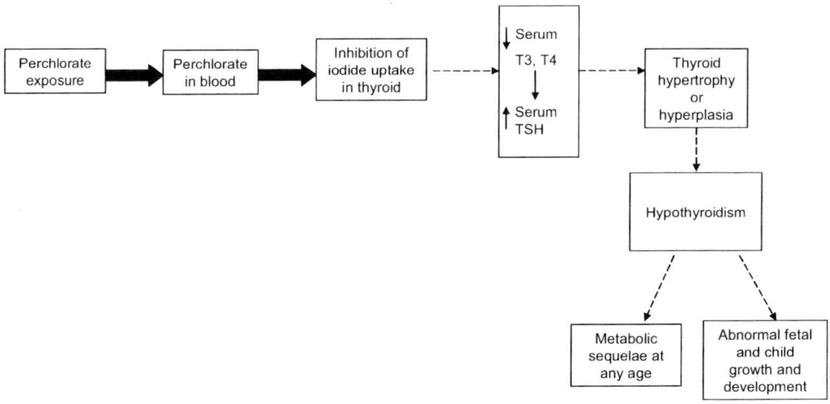

FIGURE S-1 The committee's suggested mode-of-action model of perchlorate toxicity in humans. Solid arrows represent outcomes observed in humans during perchlorate exposure. Dashed arrows represent outcomes not clearly demonstrated in humans exposed to perchlorate but biologically plausible in absence of adequate compensation. The thyroid response to increased serum TSH concentrations and the independent increase in thyroid iodide uptake would act to raise T_3 and T_4 production to normal and thus prevent later steps of the model from occurring.

risk assessment. Typically, a no-observed-adverse-effect level (NOAEL) or a lowest-observed-adverse-effect level (LOAEL) is identified from the critical study on which the RfD can be based. More recently, mathematical modeling of the dose-response data in the study has been used to provide a benchmark dose (BMD) on which the RfD can be based. The final step in the RfD process is the application of uncertainty factors to the NOAEL, LOAEL, or BMD to extrapolate from the study population to the general human population, which includes sensitive groups.

For the perchlorate risk assessment, EPA based its point of departure on reported changes in brain morphometry, thyroid histopathology, and serum thyroid hormone concentrations after oral administration of perchlorate to rats. The committee does not think that the animal data or the outcomes selected by EPA should be used as the basis of the perchlorate risk assessment. Rather, the committee recommends that inhibition of iodide uptake by the thyroid in humans, which is the key biochemical event and not an adverse effect, should be used as the basis of the risk assessment. Inhibition of iodide uptake is a more reliable and valid measure, it has been unequivocally demonstrated in humans exposed to perchlorate, and it is the key event that precedes all thyroid-mediated effects of perchlorate exposure.

The committee emphasizes that its recommendation differs from the traditional approach to deriving the RfD. The committee is recommending using a *nonadverse* effect rather than an *adverse* effect as the point of departure for the perchlorate risk assessment. Using a nonadverse effect that is upstream of the adverse effects is a conservative, health-protective approach to the perchlorate risk assessment.

The committee reviewed the human and animal data and found that the human data provided a more reliable point of departure for the risk assessment than the animal data. The committee recommends using clinical data collected in a controlled setting with the relevant route of exposure to derive the RfD. Although the data from epidemiologic studies of the general population provide some information on possible effects of perchlorate exposure, those studies are ecologic and inherently limited with respect to establishing causality and serving as a basis of quantitative risk assessment. Furthermore, those studies typically focused on changes in serum thyroid hormone and TSH concentrations or clinical manifestations of the changes, not on inhibition of iodide uptake by the thyroid. Therefore, the committee is not recommending using the available epidemiologic studies to derive the point of departure for the risk assessment. Instead, the committee recommends using the Greer et al. (2002) study in which groups of healthy men and women were administered perchlorate at 0.007-0.5 mg/kg per day for 14 days.[3] The study identified a no-observed-effect level (NOEL) for inhibition of iodide uptake by the thyroid at 0.007 mg/kg per day. The committee concludes that using the NOEL for iodide uptake inhibition from Greer et al. (2002) as the point of departure provides a reasonable and transparent approach to the perchlorate risk assessment. The NOEL value from Greer et al. (2002) is consistent with other clinical studies that have investigated iodide uptake inhibition by perchlorate. That the NOEL value from Greer et al. (2002) is a health-protective and conservative point of departure is supported by the results of a 6-month study of 0.007 mg/kg per day in a small group of healthy subjects, a 4-week study of higher doses in healthy subjects, the studies of perchlorate treatment of patients with hyperthyroidism, and extensive human and animal data that demonstrate that there will be no progression to adverse effects if no inhibition of iodide uptake occurs (see Figure S-1).

If the committee's recommendation is used as the point of departure, it

---

[3]Greer, M.A., G. Goodman, R.C. Pleus, and S.E. Greer. 2002. Health effects assessment for environmental perchlorate contamination: The dose response for inhibition of thyroidal radioiodide uptake in humans. Environ. Health Perspect. 110:927-937.

recommends using a total uncertainty factor of 10. A full factor of 10 should be used for the intraspecies factor to protect the most sensitive population—the fetuses of pregnant women who might have hypothyroidism or iodide deficiency. No additional factors are needed for duration or database uncertainties.[4] First, if inhibition of iodide uptake by the thyroid is used, chronic exposure will have no greater effect than that resulting from short-term exposure. In fact, it may well have less effect because of the capacity of the pituitary-thyroid system to compensate for iodide deficiency by increasing iodide uptake. Second, the database is sufficient, given the point of departure selected—one based on inhibition of iodide uptake by the thyroid.

---

[4]One committee member thought that the factor for database uncertainty should be greater than 1 and provided the following rationale:

> *The RfD is derived from a study in which a group of only seven healthy adults was given 0.007 mg/kg of perchlorate daily for 14 days (Greer et al. 2002). Although two other studies had similar results, the total number of subjects is still small. In addition to the small number of subjects, no chronic exposure studies have been published. An uncertainty factor of 3 could account for the uncertainty surrounding the small number of subjects and the absence of a long-term study.*

The other committee members provided the following response:

> *Although the committee acknowledges that the low-dose group (0.007 mg/kg per day) in Greer et al. (2002) had only seven subjects, the study examined the effects of four doses in a total of 37 subjects. In addition to the Greer et al. (2002) study, there are four other studies in which healthy adults were given perchlorate. The results of all the studies are remarkably similar (see Chapter 2, pp. 65-67). In addition to those studies, the studies of long-term treatment of hyperthyroidism and the studies of occupational and environmental exposure add confidence to the overall database. The issue concerning the absence of a long-term study is discussed in the section Subchronic-to-Chronic Extrapolation Factor in Chapter 5. Briefly, the key is recognizing that the committee is recommending that the RfD be based on inhibition of iodide uptake by the thyroid, a non-adverse biochemical event that precedes any adverse effects in the mode-of-action model. If that effect is used to derive the RfD, chronic exposure will have no greater effect than that resulting from short-term exposure, and in fact, it may well have less effect because of the capacity of the pituitary-thyroid system to compensate for iodide deficiency by increasing iodide uptake (see Chapter 5, p. 175).*

The committee recognizes that its recommendations would lead to an RfD of 0.0007 mg/kg per day.[5] That value is supported by other clinical studies, occupational and environmental epidemiologic studies, and studies of long-term perchlorate administration to patients with hyperthyroidism. The committee concludes that an RfD of 0.0007 mg/kg per day should protect the health of even the most sensitive populations. The committee acknowledges that the RfD may need to be adjusted upward or downward on the basis of future research, such as that suggested in this report.

RESEARCH RECOMMENDATIONS

The committee was asked to suggest scientific research that could reduce the uncertainty in the understanding of human health effects associated with perchlorate ingestion at low concentrations, especially research that could clarify "safe" exposure for sensitive populations. Although the committee found that available data are sufficient to derive an RfD for perchlorate, new research could provide a more complete understanding of the array of effects of perchlorate, especially regarding the effects of chronic exposure and the effects on sensitive populations. Therefore, the committee recommends a series of interrelated clinical, mechanistic, and epidemiologic studies that have the potential to define more precisely "safe" perchlorate exposures.

The committee recommends a clinical study designed to provide information on the potential chronic effects of low-dose perchlorate exposure on thyroid function, with a special focus on the ability and mechanisms of thyroid compensation. If long-term studies of humans are not possible, chronic studies in nonhuman primates could provide useful information. Studies of pregnant monkeys could also provide useful information on the effects of perchlorate on fetal and neonatal development. Further toxicology studies of perchlorate in rats would be less useful for clarifying the health effects of perchlorate in humans.

Especially critical issues in perchlorate risk assessment have been the effect of perchlorate on placental and breast iodide transport and the influence of iodide status on the effects of perchlorate. The committee recommends a series of in vitro studies using human tissues and animal studies to determine the role of NIS in placental iodide transport, the susceptibility of

---

[5]For comparison, EPA's draft RfD in its 2002 draft risk assessment was 0.00003 mg/kg per day.

breast NIS to perchlorate inhibition, the role of iodide status in these effects, and the effects of perchlorate on development independently of effects on iodide transport. The committee notes that other tissues contain NIS, such as the salivary glands, gastric mucosa, and perhaps the choroid plexus. Studies of NIS in those tissues, and possible effects of perchlorate on them, might be done but at a much lower priority than studies of the placenta and mammary gland.

The primary sources of uncertainty in estimating an RfD for perchlorate in drinking water arise from the absence of data on possible effects of exposure among populations at greatest risk of adverse effects of iodide deficiency (pregnant women and their fetuses and newborns). Therefore, new epidemiologic research should assess the possible health effects of perchlorate exposure in those populations. Future epidemiologic research should focus on additional analyses of existing data, new studies of health effects in selected populations, and monitoring of frequencies of specific conditions in communities affected by efforts to reduce perchlorate in drinking water.

Finally, in its deliberations on the health effects of perchlorate in drinking water, the committee considered pregnant women and their fetuses to be particularly sensitive populations. Although iodide deficiency is believed to be rare in the United States, some pregnant women may have a low iodide intake. The committee believes that further research is needed to measure more precisely the extent of, and risk factors for, iodide deficiency, particularly in pregnant women and their offspring. However, while studies are being conducted, the committee emphasizes the importance of ensuring that all pregnant women have adequate iodide intake and, as a first step, recommends that consideration be given to adding iodide to all prenatal vitamins.

1

Introduction

OVER 11 million people have perchlorate in their public drinking-water supplies at concentrations of 4 ppb (4 μg/L) or higher (EPA 2004a).[1] There is no federal drinking-water standard for perchlorate, and the concentration at which a standard should be set to protect public health is being debated. EPA has the responsibility to protect the nation's drinking water and has issued draft risk assessments that provide reference doses (RfDs) that could be used to set a federal drinking-water standard. However, EPA has been criticized that it did not appropriately consider all the relevant data for its assessments and that it based its conclusions on flawed scientific studies.

Because of the controversy surrounding the concentration at which perchlorate should be regulated, EPA, the Department of Defense (DOD), the Department of Energy (DOE), and the National Aeronautics and Space Administration (NASA) asked the National Research Council (NRC) to assess the adverse health effects of perchlorate ingestion from clinical, toxicologic, medical, and public-health perspectives. They also asked the NRC to evaluate the scientific literature, including human and animal data, and to assess the key studies underlying EPA's 2002 draft risk assessment, *Perchlorate Environmental Contamination: Toxicological Review and Risk Characterization*, with respect to quality, reliability, and relevance for drawing conclusions about the health implications of exposure to low concentrations of perchlorate in drinking water. In response to the request,

[1]The estimate of 11 million people is based on sampling data collected as of May 2004 by the U.S. Environmental Protection Agency (EPA) as required by the Unregulated Contaminant Monitoring Rule. The minimum reporting level for data collection under the Unregulated Contaminant Monitoring Rule is 4 parts per billion (ppb) (4 micrograms per liter [μg/L]).

NRC convened the Committee to Assess the Health Implications of Perchlorate Ingestion, which prepared this report.

REGULATORY HISTORY

In 1985, the Region 9 Office of EPA raised concern about potential perchlorate contamination at Superfund sites in the San Gabriel Valley in California (see Figure 1-1 for timeline of selected perchlorate-related regulatory activities) (Takata 1985). No validated analytic method was available to measure low perchlorate concentrations, and little information on their possible health effects was available (EPA 2002a). As a result, attention was focused on other chemicals at the California sites.

In the early 1990s, perchlorate contamination in monitoring wells at a California Superfund site was confirmed at concentrations greater than 1 part per million (ppm) (1 milligram per liter [mg/L]), and a provisional RfD was issued by the EPA Superfund Technical Support Center in 1992 (EPA 2002a). A revised provisional RfD was released in 1995. The RfDs were considered provisional because they had not undergone internal or external peer review. However, they were used to derive guidance levels for groundwater remediation (see Table 1-1).

In March 1997, Toxicology Excellence for Risk Assessment (TERA), a nonprofit risk assessment consulting firm, convened an independent peer review to evaluate an RfD that it had derived for perchlorate (EPA 2002a). The peer review concluded that the scientific database was insufficient to conduct a "credible quantitative risk analysis." As a result, an independent peer-review panel met in May 1997 and developed a testing strategy to address data gaps and reduce uncertainties regarding possible health effects of low-concentration perchlorate ingestion. The panel recommended a subchronic oral bioassay in rats, a developmental-neurotoxicity study in rats, a developmental study in rabbits, a two-generation reproductive toxicity study in rats, pharmacokinetic and mechanistic studies in test animals and humans, and genotoxicity and immunotoxicity assays.

In 1998, perchlorate was placed on EPA's final version of the Contaminant Candidate List (CCL), which names unregulated contaminants that may pose a public-health concern in drinking water (EPA 2004b). Contaminants on the CCL are being considered for regulation; that is, they are not subjects of federal drinking-water standards. To determine the extent of perchlorate contamination of the national drinking-water supply, monitoring of perchlorate in all large public water systems and a representative sample of small systems became mandatory beginning in 2001 (EPA 2004c).

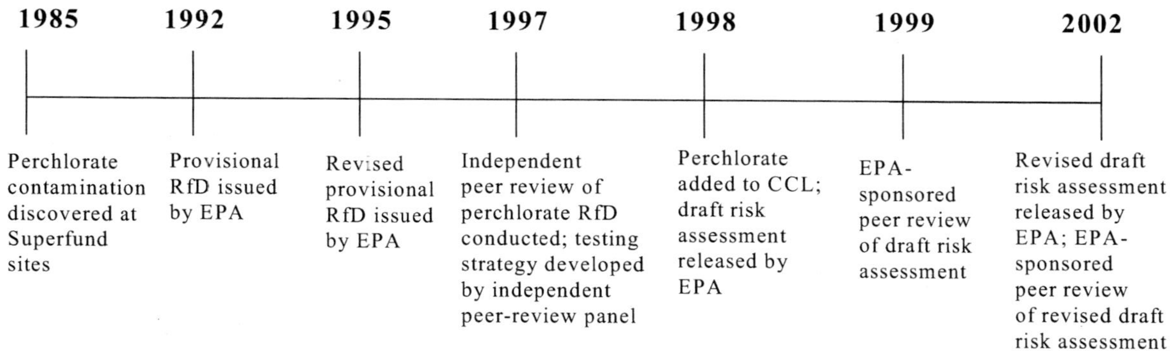

FIGURE 1-1 Timeline of perchlorate-related regulatory activities.

Although all the studies proposed by the 1997 independent peer-review panel had not been completed, EPA released its first formal draft risk assessment of perchlorate in December 1998. An EPA-sponsored peer review of the risk assessment was convened in February 1999. The panel suggested completion of the studies recommended earlier, conduct of a few additional studies to evaluate further the effects of perchlorate on fetal development, and review of the thyroid histopathology data reported in several studies by an independent group (RTI 1999). EPA issued a revised draft risk assessment in 2002 that incorporated revisions suggested by the peer review and new data generated as of fall 2001. It convened a new peer-review panel in March 2002 to review the revised risk assessment. Comments and suggestions made by that panel are being addressed by EPA. Table 1-1 lists the RfDs proposed by EPA and the corresponding drinking-water guidelines that could be derived from them using standard assumptions regarding body weight and water consumption.

CHARACTERISTICS OF PERCHLORATE

Perchlorate is a negatively charged ion (an anion) that is composed of one chlorine atom and four oxygen atoms (ClO_4^-). It is a poor complexing agent and forms a weak association with its counterion (a positively charged ion, or cation). Accordingly, perchlorate salts are extremely soluble in aqueous media and polar organic solvents. The order of solubility of the more common perchlorate salts is sodium > lithium > ammonium > potassium (Mendiratta et al. 1996). Because those perchlorate salts are so soluble, the health risks associated with them are considered equivalent to those associated with perchlorate itself, and the terms "perchlorate," "perchlorate salts," and "perchlorates" are often used interchangeably in the risk-assessment literature.

Perchlorate has excellent oxidizing ability under some conditions (EPA 2002a). However, the activation energy required to initiate the chemical reaction is very high. The high activation energy and solubility of the salts lead to perchlorate's stability and mobility in the environment. The high activation energy also leads to perchlorate's nonreactivity in the human body, where it is excreted virtually unchanged as indicated by absorption, distribution, metabolism, and elimination studies.

The primary exposure pathway of concern for perchlorate is ingestion because of its rapid uptake from the gastrointestinal tract (EPA 2002a). Dermal uptake is minimal, and the low vapor pressure of the salts leads to negligible inhalation. People might be exposed to perchlorate dust or

TABLE 1-1 EPA Provisional or Proposed Reference Doses (RfDs) and Corresponding Drinking-Water Concentrations[a]

Provisional or Proposed RfD (mg/kg per day)	Corresponding Drinking-Water Concentration (ppb)[b]	Publication Date
0.0001	4[c]	1992
0.0001-0.0005	4-18	1995
0.0009	32	1998
0.00003	1	2002

[a]Drinking-water concentrations derived from standard assumptions about body weight (70 kg) and water consumption (2 L/day).
[b]1 ppb = 1 µg/L.
[c]Example calculation: [(0.0001 mg/kg per day × 70 kg) / 2 L per day] × 1,000 µg/mg = 4 µg/L (4 ppb).

particles primarily in an occupational setting. The risk posed by that exposure would depend on the particle size distribution, which determines whether a particle is inhalable and, if it is inhalable, where in the respiratory tract it is deposited, which might affect solubility and absorption. Thus, the major route of concern is ingestion.

USE AND OCCURRENCE OF PERCHLORATE

The outstanding oxidizing ability of perchlorate led to its early use as a propellant and an explosive (Mendiratta et al. 1996). France, Germany, Switzerland, and the United States began production in the 1890s. Before the 1940s, annual global production of perchlorate was estimated to be 1,800 tons. In the middle 1940s, annual perchlorate production increased dramatically to 18,000 tons because of demand by the military and aerospace industry. Current production values are difficult to estimate because ammonium perchlorate is classified as a strategic compound.

Perchlorate is used primarily as an oxidizer in solid rocket fuels and propellants (Mendiratta et al. 1996). Ammonium perchlorate is the perchlorate salt most commonly used for that purpose. Perchlorate is also used in explosives, pyrotechnics, and blasting formulations. Magnesium perchlorate and lithium perchlorate are used in dry batteries. Other uses have been reported (EPA 2002a; Mendiratta et al. 1996).

Over the past 50 years, perchlorate has been used to diagnose and treat thyroid disease. It was used in the 1950s and 1960s to treat hyperthyroidism associated with Graves disease (EPA 2002a). The extent of its use was rather limited, and its use was curtailed when severe hematologic side effects (aplastic anemia and agranulocytosis) were reported and better antithyroid drugs became available. Today, perchlorate is used diagnostically to detect defects in the synthesis of thyroid hormones (Meier and Burger 2000) . It is also used as a treatment for patients who have developed hyperthyroidism after exposure to the antiarrythmic drug amiodarone (Martino et al. 2001). However, the Food and Drug Administration (FDA) does not recognize perchlorate as a pharmaceutical to treat endocrine or metabolic disorders (D. Orloff, Division of Metabolic and Endocrine Drug Products, FDA, personal commun., January 2004), and it is rarely used to treat any type of hyperthyroidism in the United States.

As of September 2004, environmental perchlorate releases have been confirmed in 35 states (EPA 2004d). The presence of perchlorate in the environment is reported to be associated primarily with the manufacture or use of perchlorates in solid rocket fuels and propellants (EPA 2002a). Environmental releases have also been associated with explosives and fireworks manufacture and disposal. There has been some debate about fertilizers as potential sources of perchlorate contamination (TRC 1998; Susarla et al. 1999, 2000). However, EPA (2001) conducted a survey of fertilizer composition and detected perchlorate only in products derived from Chilean caliche, an ore containing nitrates. EPA concluded that fertilizer use would probably not be a major source of perchlorate contamination and would be possible only where fertilizers derived from Chilean caliche were used.

Monitoring of perchlorate in all large public water systems and a representative sample of small systems began in 2001 (EPA 2004a,c). Data as of May 2004 indicate that perchlorate in public drinking-water supplies ranges from less than 4 ppb (minimum reporting level) to 200 ppb, with 6.4 ppb as the median concentration of values above the minimum reporting level (see Table 1-2). The highest concentration reported in the survey thus far is 420 ppb in a water facility in Puerto Rico. Data on California indicate that perchlorate ranges from less than 4 to 67 ppb, with 6.7 ppb as the median concentration of values above the minimum reporting level. The data do not represent perchlorate concentrations in all water sources. Higher perchlorate concentrations have been noted in monitoring wells associated with Superfund sites and other groundwater and surface water not directly associated with drinking-water supplies (EPA 2002a). As noted previously, the data include only a representative sample of small water facilities.

TABLE 1-2 Perchlorate Drinking-Water Concentrations Based on Monitoring Data from Unregulated Contaminant Monitoring Rule as of May 2004

	Perchlorate			Samples		
Location	Minimal Concentration (ppb)[a]	Maximal Concentration (ppb)	Median Concentration (ppb)[b]	Number above MRL	Total	Fraction below MRL (%)
United States	<4	200[c]	6.4	535	28,179	98.1
Puerto Rico	<4	420	NA	1	734	99.9
California	<4	67[d]	6.7	364	8,179	95.5

[a]1 ppb = 1 μg/L.
[b]Median perchlorate concentration of values above the minimum reporting level (<4 ppb).
[c]Maximum in United States was measured in Duval County, Florida.
[d]Maximum in California was measured in San Bernardino County.
Abbreviations: MRL, minimum reporting level (<4 ppb); NA, not applicable.
Source: Data from EPA 2004a.

Information presented in a recent report from the University of California, Irvine, indicates much higher perchlorate concentrations in California drinking water (Bull et al. 2004). However, that report includes data from 1997 to 2004 and, more important, data on inactive, abandoned, and destroyed water sources. The EPA data (2004a) are only on active drinking-water sources, and this could account for the differences in the two sets of data.

Perchlorate has also been detected in food sources. For example, the Environmental Working Group, a nonprofit research organization, reported that perchlorate was detected in four of 22 samples of commercial lettuce purchased in January and February 2003 from seven grocery stores in northern California (Sharp and Lunder 2003). Perchlorate concentrations in the four lettuce samples ranged from 30 to 121 ppb (nanograms per gram of lettuce wet weight); the average was 70 ppb. Perchlorate has been detected in commercial milk samples. Kirk et al. (2003) found perchlorate in all seven samples of milk randomly purchased from seven grocery stores in Lubbock, Texas. Concentrations ranged from 1.75 to 6.30 ppb. Perchlorate also was detected in a sample of evaporated milk and a sample of breast milk but not in a sample of powdered milk. More recent surveys of milk have shown perchlorate concentrations ranging from nondetectable to 10.6 ppb (Sharp 2004). Perchlorate concentrations in breast milk have been measured in a current study in Chile (Gibbs 2004); median concentrations of 19.3 and 103.8 ppb were measured in Chanaral and Taltal, where mean perchlorate concentrations in tap water are 5.8 and 113 ppb, respectively. FDA has developed analytic techniques appropriate for measuring perchlorate in various food sources and is conducting a survey of foods to determine the extent of perchlorate contamination in suspect foods (FDA 2004). Preliminary results of sampling by FDA found perchlorate concentrations ranging from 3 to 11 ppb in 20 milk samples.

Thus, the contamination of public water supplies and food sources has raised substantial concerns about the effects of low-level perchlorate ingestion and forced the debate over a drinking-water standard into the public spotlight.

SENSITIVE POPULATIONS

Transfer of iodide from blood into the thyroid gland is essential for the synthesis of the thyroid hormones: thyroxine (T_4) and triiodothyronine (T_3) (see Chapter 2). Perchlorate blocks the transport of iodide into the thyroid gland, which could potentially lead to iodide deficiency and decreased synthesis of T_3 and T_4. The thyroid hormones are critical determinants of growth and development in fetuses, infants, and young children. Thus, fetuses and preterm newborns constitute the most sensitive populations although infants and developing children are also considered sensitive populations. People who have compromised thyroid function resulting from conditions that reduce thyroid hormone production and people who are iodide-deficient also constitute potentially sensitive populations.

SCIENTIFIC CONTROVERSIES

Several issues have been repeatedly raised at conferences and in peer reviews concerning EPA's assessment of potential adverse health effects of perchlorate exposure (EPA 2002b; Schwartz et al. 2004). The adequacy and relevance of available human data for assessing health risks posed by perchlorate exposure have been debated. Clinical, occupational, and epidemiologic data are available, and some argue that they should be used to determine the RfD, the key value used to derive the drinking-water standard. However, others state that the human studies should not be used as the basis of the RfD, because they contain one or more of the following limitations: lack of control of confounding factors, evaluation only of healthy adults, inadequate exposure information, evaluation of effects over short exposure durations, and assessment of a narrow set of toxicity end points.

The quality and validity of some animal data have also been debated (EPA 2002b; Schwartz et al. 2004). Specifically, questions have been raised about neurodevelopmental studies in which rats were exposed to perchlorate in utero and postnatally and then examined for changes in specific areas of the brain. Many have challenged the experimental methods, the statistical analysis of the data, the interpretation of the reported findings, and the inconsistencies between the reported findings and the general literature on thyroid hormones and brain development. Others have argued that those data cannot be disregarded in the assessment of potential adverse health effects of perchlorate ingestion.

Another point of contention is the definition of the adverse health effect associated with perchlorate ingestion (EPA 2002b; Schwartz et al. 2004). EPA has proposed a mode-of-action model for perchlorate toxicity, which shows a continuum of possible health effects of perchlorate exposure (see Figure 1-2). The effect that is defined as the adverse effect is a matter of

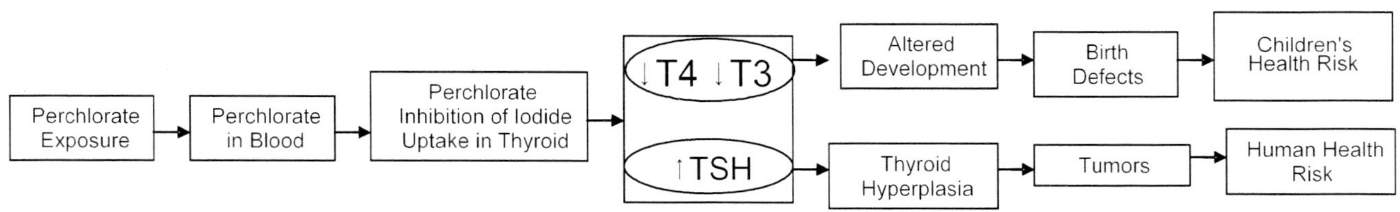

FIGURE 1-2 EPA's proposed continuum of possible health effects of perchlorate exposure. Source: EPA 2002a.

debate. Some believe that the adverse effect should be defined as inhibition of iodide uptake by the thyroid or as decreases in T_3 and T_4 production with corresponding increases in thyroid-stimulating hormone (TSH) production. Others contend that those effects are "preadaptive" or "adaptive" effects and that the adverse effect is one or more clinical manifestations of hypothyroidism, such as developmental deficits. Defining the adverse effect is important because it influences how the RfD is derived and ultimately the value of the drinking-water standard.

Finally, the application of various uncertainty factors has become a source of controversy. When an RfD is calculated, uncertainty factors are used to extrapolate from the study population to the larger general population. Those factors account for interspecies differences (extrapolation from animal to human populations, if applicable), intraspecies differences (possible variations or sensitivities that might be present in the general population), failure to identify a no-observed-adverse-effect level, absence of chronic toxicity data, and other database gaps. The uncertainty factors typically range from 1 to 10; 1, 3, and 10 are the values most commonly used. No absolute rules exist for application of the factors, and professional judgment is a large component of their use. Regarding derivation of EPA's RfD for perchlorate, some have questioned the use of specific uncertainty factors, particularly the factor used to account for database uncertainties (see Chapter 5 for further discussion of uncertainty factors). Questions regarding the use of uncertainty factors and the other issues discussed above all played a role in determining the charge provided to this committee.

COMMITTEE'S CHARGE AND APPROACH TO CHARGE

The members of the NRC committee were selected for their expertise in pediatrics; endocrinology; pediatric endocrinology; thyroid endocrinology, physiology, and carcinogenesis; immunology; veterinary pathology; animal toxicology; neurotoxicology; developmental toxicology; physiologically based pharmacokinetic modeling; epidemiology; biostatistics; and risk assessment. The committee was asked to accomplish the following tasks:

- Evaluate the current state of the science regarding potential adverse effects of disruption of thyroid function in humans and laboratory animals at various stages of life. Specifically, evaluate whether science supports the model that predicts potential adverse neurodevelopmental and neoplastic effects from changes in thyroid hormone regulation that result from disruption of iodide uptake by the thyroid gland, and indicate the level of confidence in the model.

- Assess the levels at which chronic inhibition of iodide uptake may lead to adverse (not just adaptive) health effects in humans, especially sensitive populations. Consider the influence of iodide in the diet on the levels at which adverse effects would be observed, especially in sensitive populations, and indicate the degree of confidence in the proposed levels.
- Assess the levels at which changes in thyroid hormones may lead to adverse (not just adaptive) health effects in humans, especially sensitive populations, and indicate the level of confidence in those values.
- Evaluate the animal studies used to assess human health effects of ingestion of perchlorate with particular attention to key end points, including changes in brain morphometry, behavior, thyroid hormone concentrations, and thyroid histopathology. Indicate the level of confidence in the relevance of the adverse effects observed in the animal studies to human health, especially sensitive populations, and specifically address the validity of the model that extrapolates changes in brain morphometry in rats to adverse effects in the human population, especially sensitive populations. Indicate whether adverse effects, other than those associated with iodide uptake inhibition, may result from daily ingestion of perchlorate at low concentrations.
- On the basis of the above review, determine whether EPA's findings in its 2002 draft risk assessment, *Perchlorate Environmental Contamination: Toxicological Review and Risk Characterization*, are consistent with current scientific evidence. Specifically, determine whether EPA considered all relevant literature (both supporting and nonsupporting), consistently critiqued that literature, and then used appropriate scientific studies to develop its health risk assessment. If deficiencies in EPA's analysis are found, such as lack of consideration of a key study, provide suggestions as to how EPA might modify its assessment. Provide a range of values consistent with scientific evidence for percentage iodide uptake that would protect persons at various life stages and with varied thyroid status.[2] On the basis of the scientific evidence, provide information that can be used to inform the selection of uncertainty factors used in the approximation of a safe lifetime perchlorate exposure for humans, especially sensitive populations.
- Finally, suggest specific scientific research that could reduce the uncertainty in the current understanding of human health effects associated

---

[2]The committee interpreted this task as referring to iodide intake that would protect people at various life stages and with varied thyroid status.

with low-level perchlorate ingestion. Specifically, suggest research that could clarify safe levels of exposure of sensitive populations, and provide rough estimates of timeframe, costs, and potential to reduce overall uncertainty for the specific research studies suggested.

To accomplish its task, the committee held five meetings from October 2003 to July 2004. Public sessions were held during the first, second, and fourth meetings, in which the committee heard presentations from representatives of the Office of Science and Technology Policy, EPA, DOD, DOE, NASA, the Food and Drug Administration, Congress, the Agency for Toxic Substances and Disease Registry, California EPA, and other interested parties, including industry and environmental groups (see Appendix C). In the second meeting, several noted scientists were invited to make presentations to the committee to answer questions raised in the first meeting. The committee reviewed (1) materials submitted by EPA, DOD, DOE, NASA, industry, and private individuals, (2) studies evaluated in EPA's 2002 draft perchlorate risk assessment, (3) findings in EPA's 2002 draft perchlorate risk assessment, and (4) information from publicly available scientific literature. Accordingly, the committee evaluated both published and unpublished data; however, it typically gave more weight in its deliberations to published reports. Unpublished data were considered only when the committee had sufficient information to evaluate the methods used to produce them. Overall, emphasis was given to studies with the soundest scientific methods to draw conclusions regarding effects of perchlorate exposure.

For each proposed adverse health effect of perchlorate exposure, the committee evaluated all the evidence gathered. Conclusions were based on the following categories: (1) no evidence, (2) evidence is inadequate to accept or reject a causal relationship, (3) evidence favors rejection of a causal relationship, (4) evidence favors acceptance of a causal relationship, and (5) evidence establishes a causal relationship. Using that hierarchy where appropriate, the committee was able to reach conclusions regarding the potential adverse effects of perchlorate exposure.

ORGANIZATION OF REPORT

This report is divided into six chapters. Chapter 2 provides information on thyroid function in humans and possible effects of disruption of thyroid function in adults, neonates, and fetuses; it also discusses clinical studies in which humans were exposed to perchlorate. Chapter 3 reviews the epidemiologic studies of occupational and environmental exposure to perchlorate, including strengths and weaknesses of the studies. Chapter 4 reviews animal toxicity studies with emphasis on studies of the effect of perchlorate on thyroid hormone production, thyroid histopathology, brain morphometry, neurobehavior, and thyroid tumors. Chapter 4 also addresses the possibility of effects of perchlorate exposure that are independent of inhibition of iodide uptake by the thyroid. Chapter 5 reviews the committee's critique of EPA's 2002 draft risk assessment, *Perchlorate Environmental Contamination: Toxicological Review and Risk Characterization.* Research that could reduce uncertainty regarding health effects of perchlorate exposure is presented in Chapter 6.

REFERENCES

Bull, R.J., A.C. Chang, C.F. Cranor, R.C. Shank, and R. Trussell. 2004. Perchlorate in Drinking Water: A Science and Policy Review. Urban Water Research Center, University of California, Irvine, CA. [Online]. Available: http://-www.urbanwater.uci.edu/UCI-UWRC_Perchlorate_wCorrection061404.pdf [accessed August 18, 2004]

EPA (U.S. Environmental Protection Agency). 2001. Survey of Fertilizers and Related Materials for Perchlorate (ClO4): Final Report. EPA/600/R-01/049. Office of Research and Development, U.S. Environmental Protection Agency, Cincinnati, OH. [Online]. Available: http://www.epa.gov/ORD/NRMRL/Pubs/600/R01/047.pdf [accessed August 18, 2004].

EPA (U.S. Environmental Protection Agency). 2002a. Perchlorate Environmental Contamination: Toxicological Review and Risk Characterization. External Review Draft. NCEA-1-0503. National Center for Environmental Assessment, Office of Research and Development, U.S. Environmental Protection Agency, Washington, DC.

EPA (U.S. Environmental Protection Agency). 2002b. Report on the Peer Review of the U.S. Environmental Protection Agency's Draft External Review Document "Perchlorate Environmental Contamination: Toxicological Review and Risk Characterization." EPA/635/R-02/003. National Center for Environmental Assessment, Office of Research and Development, U.S. Environmental Protection Agency, Washington, DC. [Online]. Available: http://www.epa.gov/ncea/pdfs/perchlorate/final_rpt.pdf [accessed August 18, 2004].

EPA (U.S. Environmental Protection Agency). 2004a. Unregulated Contaminant Monitoring Rule (UCMR) 1999. UCMR Data.. Office of Ground Water and Drinking Water, U.S. Environmental Protection Agency. [Online]. Available: http://www.epa.gov/safewater/ucmr.html [accessed June 2004].

EPA (U.S. Environmental Protection Agency). 2004b. Drinking Water Contaminant Candidate List. Office of Ground Water and Drinking Water, U.S. Environmental Protection Agency. [Online]. Available: http://www.epa.gov/safewater/ccl/cclfs.htm [accessed March 2004].

EPA (U.S. Environmental Protection Agency). 2004c. Revisions to the Unregulated Contaminant Monitoring Rule Fact Sheet. EPA 815-F-01-008. Office of Water, U.S. Environmental Protection Agency. [Online]. Available: http://www.epa.gov/safewater/standard/ucmr/ucmrfact.html [accessed March 2004].

EPA (U.S. Environmental Protection Agency). 2004d. Known Perchlorate Releases in the U.S. - September 23, 2004. Perchlorate Occurrences. Federal Facilities Restoration and Reuse Office, Office of Solid Waste and Emergency Response, U.S. Environmental Protection Agency [Online]. Available: http://www.epa.gov/fedfac/detection_with_dates09_23_04.xls [accessed Nov. 15, 2004].

FDA (Food and Drug Administration). 2004. Perchlorate: Questions and Answers. Center for Food Safety and Applied Nutrition, U.S. Food and Drug Administration. September 20, 2003. [Online]. Available: http://www.cfsan.fda.gov/~dms/clo4qa.html [accessed June 2004].

Gibbs, J.P. 2004. Chronic Environmental Exposure to Perchlorate in Drinking Water and Thyroid Function during Pregnancy and the Neonatal Period. Presentation at the Fourth Meeting on Assess the Health Implications of Perchlorate Ingestion, May 24, 2004, Woods Hole, MA.

Kirk, A.B., E.E. Smith, K. Tain, T.A. Anderson, and P.K. Dasgupta. 2003. Perchlorate in milk. Environ. Sci. Technol. 37(21):4979-4981.

Martino, E., L. Bartalena, F. Bogazzi, and L.E. Braverman. 2001. The effects of amiodarone on the thyroid. Endocr. Rev. 22(2):240-254.

Meier, C.A., and A.G. Burger. 2000. Effects of drugs and other substances on thyroid hormone synthesis and metabolism. Pp. 265-280 in Werner and Ingbar's The Thyroid: A Fundamental and Clinical Text, 8th Ed., L.E. Braverman, and R.D. Utiger, eds. New York: Lippincott Williams and Wilkins.

Mendiratta, S.K., R.L. Dotson, and R.T. Brooker. 1996. Perchloric acid and perchlorates. Pp. 157-170 in Kirk-Othmer Encyclopedia of Chemical Technology, 4th Ed., Vol. 18, J.I. Kroschwitz, and M. Howe-Grant, eds. New York: John Wiley and Sons.

RTI (Research Triangle Institute). 1999. Perchlorate Peer Review Workshop Report. EPA Contract Number 68-W98-085. Prepared for Office of Solid Waste, U.S. Environmental Protection Agency, Washington, DC, by Center for Environmental Analysis, Research Triangle Institute, Research Triangle Park, NC.

Schwartz, H.L., M. Aschner, A.J. Elberger, J.C. Lind, F.R. Brush, M.K. Cowles, A.J. DeRoos, K.C. Donnelly, J.M. Hershman, R. Wilson, J.T. Lane, D.B. Bylund, D.W. Cragin, S.C. Lewis, J.D. Wilson, and R. Wilson. 2004. Perchlorate State of the Science Symposium 2003: Report of the Planning Committee and Reports of the Expert Review Panels. Reports compiled from the

Perchlorate State of the Science Symposium, September 29-October 1, 2003, Omaha, NE.

Sharp, R. 2004. Rocket Fuel Contamination in California Milk. Environmental Working Group. [Online]. Available: http://www.ewg.org/reports/rocketmilk/ [accessed August 31, 2004].

Sharp, R., and S. Lunder. 2003. Suspect Salads: Toxic Rocket Fuel Found in Samples of Winter Lettuce, Part 1. Environmental Working Group, Washington, DC. [Online]. Available: http://www.ewg.org/reports/suspectsalads/part1.php [accessed August 18, 2004].

Susarla, S., T.W. Collette, A.W. Garrison, N.L. Wolfe, and S.C. McCutcheon. 1999. Perchlorate identification in fertilizers. Environ. Sci. Technol. 33(19): 3469-3472.

Susarla, S., T.W. Collette, A.W. Garrison, N.L. Wolfe, and S.C. McCutcheon. 2000. Perchlorate identification in fertilizers. Environ. Sci. Technol. 34(1):224.

Takata, K. 1985. Request for CDC Assistance Regarding Potential Health Effects of Perchlorate Contamination at the San Gabriel Valley Superfund Sites. Memorandum to Don Hawkins, CDC Regional Representative (T-1), from Keith Tanaka, Chief, Superfund Programs Branch (T-4), U.S. Environmental Protection Agency, Francisco, CA. December 23, 1985.

TRC (TRC Environmental Corporation). 1998. Chemical Fertilizer as a Potential Source of Perchlorate. Prepared for Lockheed Martin Corp., Burbank, CA, by TRC, Irvine, CA. November 1998.

2

The Thyroid and Disruption of Thyroid Function in Humans

THYROID hormones are critical determinants of growth and development in infants and of metabolic activity in infants and adults. They affect the functions of virtually every organ system, and they must be constantly available to carry out these functions. A steady supply of thyroid hormones is provided by large reservoirs in the circulation and the thyroid gland. Thyroid hormone biosynthesis and secretion are normally maintained within narrow limits by regulatory mechanisms that are sensitive to small changes in the circulating hormone concentrations. The regulatory mechanisms protect against both hypothyroidism (deficient thyroid hormone production) and hyperthyroidism (excess thyroid hormone production).

There are two thyroid hormones, thyroxine (T_4) and triiodothyronine (T_3).[1] All the circulating T_4 and about 20% of the T_3 are produced in the thyroid gland; the remaining 80% of the T_3 is produced from T_4 in most tissues of the body. Through a process known as deiodination, enzymes called deiodinases remove a single iodine atom from the T_4 molecule to produce the more active hormone T_3. Most of the effects of thyroid hormones in individual cells of the body are exerted by T_3, acting to regulate thyroid hormone-responsive expression of genes that code for many cellular proteins that regulate development, growth, and metabolism.

Topics discussed in this chapter include the production of T_4 and T_3, the

[1] A third hormone of thyroid origin, calcitonin, is produced by different cells within the thyroid gland. It affects calcium metabolism but has no effects on iodide metabolism or T_4 or T_3 production or action and is not discussed further in this report.

regulation of thyroid hormone production, the actions of the thyroid hormones, the development of thyroid function during fetal life, and the effects of perchlorate when given deliberately to humans. The effects of environmental exposure to perchlorate are summarized in Chapter 3.

ANATOMY

The thyroid is a butterfly-shaped gland in the front of the neck. It weighs about 1-1.5 g at birth and 10-20 g in healthy adults in the United States. It contains millions of spherical follicles, each composed of a single layer of cells known as thyroid follicular cells surrounding a space, or lumen, filled with fluid known as colloid (Figure 2-1). Thyroid follicles are the functional units of the thyroid gland. The colloid consists mostly of thyroglobulin, a thyroid protein that serves as the framework for production of T_4 and T_3 and as the storage form of the two hormones.

THYROID HORMONE PRODUCTION, TRANSPORT, AND ACTION

T_4 and T_3 are the only biologically active substances that contain iodine. They are similar in that each has two six-member rings—an inner and an outer ring (Figure 2-2)—connected by an ether linkage. Their inner rings have two iodine atoms; T_4 has two iodine atoms in its outer ring, whereas T_3 has only one. The compound formed if an iodine atom is removed from the inner ring of T_4 is 3,3',5'-triiodothyronine (reverse T_3), which has no biologic activity.

Production of Thyroxine and Triiodothyronine in the Thyroid Gland

Transport of Iodide into Thyroid Cells

Iodine is an intrinsic component of T_4 and T_3. Transfer of iodide from the circulation into the thyroid gland is therefore an essential step in the synthesis of the two hormones.[2] Iodide is transported from the bloodstream

---

[2]Iodide is the negatively charged ion of iodine and is the form of iodine that is found in foods and in the circulation in humans.

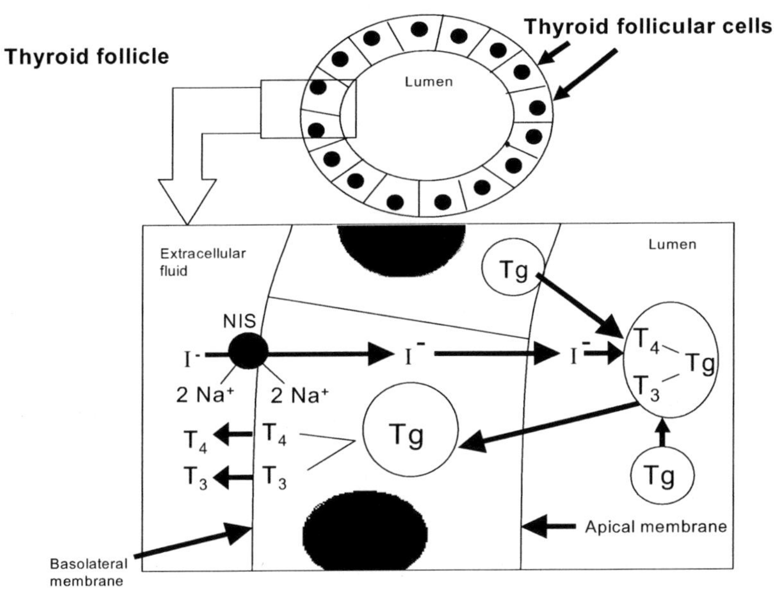

FIGURE 2-1 Diagram of thyroid cells and a thyroid follicle, showing key steps in thyroxine (T_4) and triiodothyronine (T_3) synthesis and secretion. A follicle consists of a single layer of thyroid follicular cells surrounding a lumen, which is filled with thyroglobulin (Tg). Iodide (I^-) and sodium (Na^+) ions are transported into cells via the sodium (Na^+)/iodide (I^-) symporter (NIS) in the basolateral membrane of the cells. Iodide diffuses to the luminal side of the cell and is transported into the lumen of the follicle, where it is oxidized and then used to form T_4 and T_3 (within Tg). Tg is taken up by cells and broken down, freeing its constituent T_4 and T_3 molecules, which then diffuse out of the cell and into the bloodstream.

into thyroid cells against a chemical and electric gradient. It diffuses rapidly across the cells and is transported into the lumen of thyroid follicles, where T_4 and T_3 are produced (Figure 2-1). Iodide transport into the cells is mediated by a specific protein molecule, the sodium (Na^+)/iodide (I^-) symporter (NIS) (Dohan et al. 2003). The symporter is also present in substantial quantities in the salivary glands, stomach, and mammary glands; the iodide that is transported into these tissues is not further metabolized, as it is in the thyroid gland, but instead is secreted unchanged into saliva, gastric juice, or milk. Very small amounts of the symporter have been found in other tissues (see the last section of this chapter and Chapter 4).

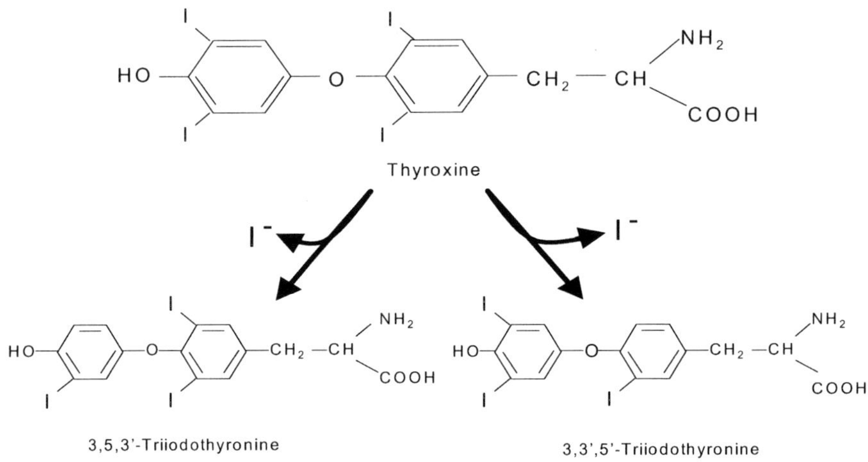

FIGURE 2-2 Structures of thyroxine (T_4), triiodothyronine (T_3), and reverse triiodothyronine (reverse T_3). Note that T_3 is missing an iodine atom on its outer ring and reverse T_3 is missing an iodine atom on its inner ring. T_3 is most potent thyroid hormone. T_4 is active only after conversion to T_3, and reverse T_3 has no biologic activity.

The NIS is a glycoprotein composed of 643 amino acids and a small amount of carbohydrate. It is in the outer (plasma) membrane at the basolateral surface of the thyroid follicular cells (Figure 2-1). It is not present in the membrane at the apical surface of the cell, the part of the cell that is adjacent to the lumen of the thyroid follicle. The symporter uses the inward movement of sodium ions to transport iodide into the cells; two sodium ions are transported for each iodide ion.

NIS has a very high affinity for iodide, which allows transport of iodide into the cells against a high concentration gradient. It also transports other ions with a similar shape and electric charge, such as perchlorate and thiocyanate. The affinity of NIS for those substances is higher than its affinity for iodide, so they can block iodide transport into thyroid cells, which can result in a decrease in the iodide concentration in the cells and therefore in a decrease in the availability of iodide for synthesis of T_4 and T_3. Among those substances, perchlorate is the best studied. It is a true competitive inhibitor of iodide transport: it blocks iodide transport through the symporter in a dose-dependent manner. Conversely, iodide blocks perchlorate transport in a similar manner. Whether perchlorate itself is transported into the thyroid gland is debated (Dohan et al. 2003; Van Sande et al. 2003).

Thyrotropin (thyroid-stimulating hormone, TSH) is produced by the anterior pituitary gland that stimulates all aspects of thyroid function, including the production of symporter molecules and therefore iodide transport. T_4, T_3, and cytokines (molecules that mediate inflammatory reactions) inhibit TSH secretion, thereby decreasing the production of symporter molecules. Iodide deficiency results in an increase in production of symporter molecules independently of TSH. A few patients have had mutations in the gene for the NIS. Because their thyroid follicular cells cannot transport iodide, they are unable to synthesize T_4 or T_3, and consequently they have hypothyroidism.

Synthesis and Secretion of Thyroxine and Triiodothyronine in the Thyroid Gland

After passage through the symporter into thyroid follicular cells, iodide rapidly diffuses to the apical surface of the cells (Figure 2-1). There, it is transported across the apical membrane into the lumen of the follicles by pendrin, a membrane iodide-chloride transporter. It is then rapidly oxidized and covalently bound ("organified") to specific tyrosine residues of thyroglobulin; some tyrosine residues gain one iodine atom (forming monoiodotyrosine), and others gain two iodine atoms (forming diiodotyrosine). These oxidation and binding reactions are catalyzed by the enzyme thyroid peroxidase in a reaction that requires hydrogen peroxide (Taurog 2000).

Thyroglobulin is a large protein that serves as the site of thyroid hormone synthesis. It is synthesized by the thyroid follicular cells and then carried to the follicular lumen. Within the thyroglobulin molecule, T_4 is formed by the coupling of two diiodotyrosine residues, and T_3 is formed by the coupling of one monoiodotyrosine residue to one diiodotyrosine residue. Those reactions also are catalyzed by thyroid peroxidase. The coupling process is not random; T_4 and T_3 are formed in regions of the thyroglobulin molecule that have unique amino acid sequences (Dunn and Dunn 2000). A normal thyroglobulin molecule contains about six molecules of monoiodotyrosine, four of diiodotyrosine, and two of T_4 or T_3.

To liberate T_4 and T_3, thyroglobulin is taken up from the lumen of the thyroid follicles into the thyroid follicular cells (Figure 2-1). It is then broken down into T_4, T_3, and its constituent amino acids. The hormones are then released into the circulation.

Extrathyroid Production of Thyroid Hormones

About 80% of the T_3 produced each day is formed by removal of one iodine atom from the outer ring of T_4 (outer-ring deiodination) in tissues outside the thyroid gland, such as the liver, kidneys, muscle, and nervous system. The deiodination reaction is stimulated (catalyzed) by enzymes called deiodinases. Two deiodinases—type 1 and type 2—catalyze the conversion of T_4 to T_3. Type 3 deiodinase catalyzes the conversion of T_4 to reverse T_3 and of T_3 to 3,3'-diiodothyronine (inner-ring deiodination); neither reverse T_3 nor 3,3'-diiodothyronine has biologic activity (Bianco et al. 2002).

The three deiodinases differ in tissue distribution and regulation. Type 1 deiodinase is the predominant deiodinating enzyme in the liver, kidneys, and thyroid. Type 2 is the predominant deiodinating enzyme in the brain, pituitary gland, heart, and muscle. Type 3 is prominent in the skin, brain, uterus, and placenta.

Thyroid Hormone Production Rate

There are major quantitative and qualitative differences between T_4 and T_3 in production and metabolism. In healthy adults, the production rate of T_4 is 80-100 μg per day, all of which is produced in the thyroid gland. T_4 is degraded at about 10% per day, mostly by deiodination to form T_3 or inactive reverse T_3. The production rate of T_3 is 30-40 μg per day, of which about 20% is produced in the thyroid gland and 80% by the extrathyroid deiodination of T_4. T_3 is degraded about 75% per day, mostly by deiodination.

Most or all of the biologic activity of thyroid hormones is exerted by T_3, whether produced in the thyroid or from T_4 in other tissues. T_4 is largely a prohormone, with little, if any, intrinsic biologic activity, and its conversion to T_3 in effect produces the active thyroid hormone. The conversion process in extrathyroidal tissues is regulated, so production of T_3 may change independently of changes in the function of the thyroid gland itself.

Transport of Thyroid Hormones in Serum

The thyroid hormones circulate in the bloodstream in two forms: some as the free (unbound) hormone and most bound to protein. Some 99.97% of the T_4 and 99.7% of the T_3 are protein-bound. The serum proteins that bind the hormones are thyroxine-binding globulin (TBG), transthyretin, albumin, and lipoproteins. TBG is the most important: it carries 75-85% of the T_4 and T_3 in serum. As a result of the binding of T_4 and T_3 to those proteins, clearance of the hormones from the circulation is slow; the serum half-life of T_4 is 5-7 days, and that of T_3 is about 20 hours (hr). Thus, even when release of T_4 and T_3 from the thyroid gland abruptly ceases—for example, when the thyroid gland is surgically removed—the serum concentrations of the two hormones fall slowly. The fall is considerably smaller and slower when T_4 and T_3 production is only partially reduced.

The binding of T_4 and T_3 to the serum proteins is not a determinant of the actions of the hormones in tissues. People with no TBG (a rare disorder caused by a mutation of the gene for TBG) or with twice the normal amount (caused by pregnancy, treatment with estrogen, or a duplication of the gene for TBG) have normal thyroid function. They have, respectively, low or high serum total T_4 and total T_3 concentrations but normal serum free (unbound) T_4 and T_3 concentrations. Those findings establish that it is the serum free T_4 and T_3 concentrations that determine the hormones' biologic activity and ultimately the clinical status of people. The binding proteins maintain the serum free T_4 and T_3 concentrations within narrow limits and ensure not only that the hormones are continuously available to tissues but also that more free hormone can be made available almost instantly from the large amounts bound to the proteins in the serum if a sudden need arises. Because T_4 and T_3 bind to the proteins so well, tissues are protected from surges in T_4 or T_3 in the circulation. The proteins thus have both storage and buffering functions.

Cellular Uptake and Actions of Thyroid Hormones in Tissues

Free T_4 and free T_3 in serum are available for uptake into cells at any time. They are carried into cells primarily by transporter molecules in the cell membranes (Hennemann et al. 2001). There are several transporters, with different affinity for T_4 and T_3, and some hormone may enter cells by passive diffusion.

T_3 is also available to cells because it is produced from T_4 in them (Figure 2-3). Some of the T_3 produced in the cells returns to the circulation, and some remains in the cells. Indeed, locally produced T_3 provides most of the T_3 found in the nuclei of the cells in many tissues. Thus, there are two sources of T_3 in cells: some enters the cells from the circulation, and

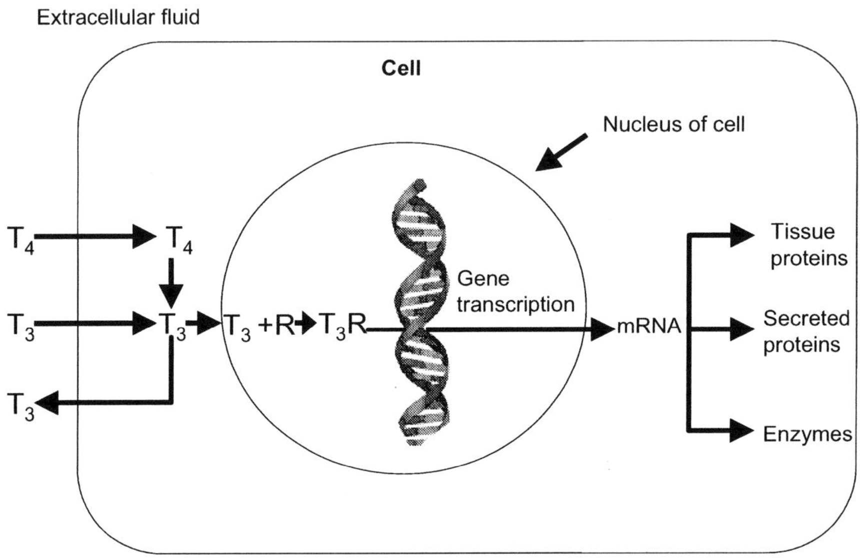

FIGURE 2-3 Diagram of a cell showing pathways of thyroxine (T_4) and triiodothyronine (T_3) metabolism. T_4 and T_3 are transported into cells, after which T_4 is converted to T_3 by action of deiodinases, thereby increasing the pool of T_3 in the cell. Some T_3 enters the nucleus of the cell, where it binds to specific T_3 receptors (R). Hormone-receptor complexes then bind to DNA and alter gene transcription and therefore synthesis of many proteins, including tissue proteins, proteins that are secreted for action elsewhere, and enzymes that regulate metabolic activity. Some T_3 also leaves the cell and enters the bloodstream. Source: American Heritage 2003. DNA illustration by Carlyn Iverson. Reprinted by permission of American Heritage Children's Science Dictionary, copyright 2003.

some is produced in the cells by deiodination of T_4. The relative contributions of the two sources to the T_3 that is bound to its receptors in the nuclei of cells vary substantially from tissue to tissue (Bianco et al. 2002).

In cells, T_3 diffuses into the nuclei, where it binds to specific nuclear receptors (Figure 2-3). The T_3-receptor complexes then bind to the regulatory regions of many different DNA molecules (genes). The interaction of the genes with the receptor-hormone complexes alters the rate at which the genes synthesize molecules of messenger RNA, and thus leads to changes in the rate of synthesis of thyroid hormone-dependent proteins (Mariash et al. 2000). T_3 may also have some actions at other sites of cells, such as the cell membrane, but the biologic importance of these actions is not known.

Some biologic responses to T_3 are stimulatory, and others inhibitory. By stimulating the production of several proteins in the heart, T_3 increases heart rate and contractility. In the liver, it stimulates the synthesis of many different proteins required for growth, metabolism, and energy production. It also stimulates the production of proteins in the brain, most obviously during development. In contrast, in the pituitary gland, T_3 inhibits the production of TSH, a process termed negative feedback, which ultimately leads to a decrease in hormone synthesis by the thyroid gland.

REGULATION OF THYROID HORMONE PRODUCTION

Thyroid hormone production is regulated in two ways:

- Regulation of thyroid gland synthesis and secretion of T_4 and T_3 by TSH. Pituitary secretion of TSH is inhibited by T_4 and T_3 and stimulated by a decrease in T_4 and T_3. TSH secretion is also stimulated by thyrotropin-releasing hormone, produced in the hypothalamus.
- Regulation of extrathyroid conversion of T_4 to T_3 by several hormones, including T_4 and T_3 themselves, and by nutritional and illness-related factors. The effect of those factors differs in different tissues; for example, starvation results in a decrease in conversion in the liver, but not in the brain.

The first mechanism provides a sensitive defense against increases and especially decreases in thyroid hormone production. The second mechanism provides for rapid changes in the availability of T_3 in different tissues, especially in response to illness or starvation.

Thyrotropin-Releasing Hormone

Thyrotropin-releasing hormone is distributed throughout the nervous system and elsewhere, where it is thought to modulate transmission of nerve impulses. Its content is highest in the hypothalamus, and the thyrotropin-releasing hormone produced in this region is an important regulator of TSH secretion and therefore of T_4 and T_3 production by the thyroid gland (Figure 2-4). The production and secretion of thyrotropin-releasing hormone in the

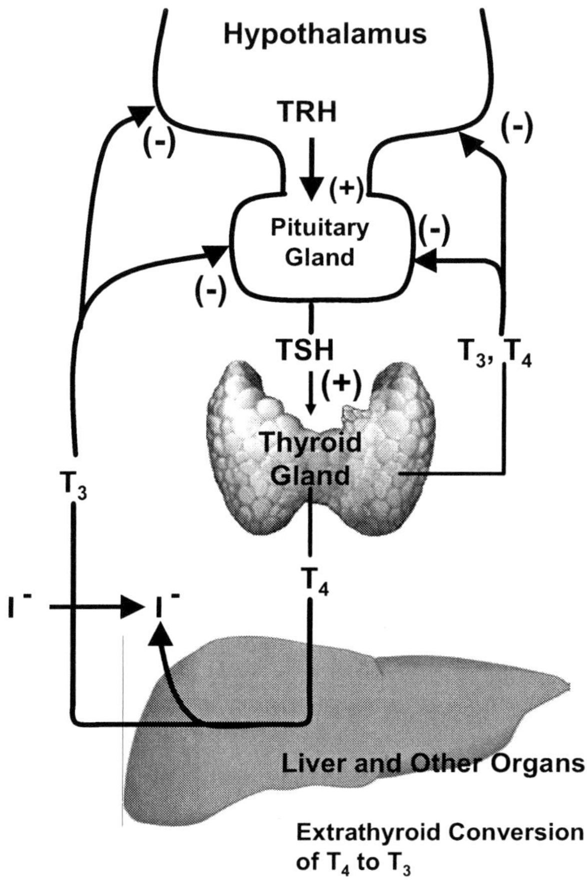

FIGURE 2-4 Diagram of the hypothalamic-pituitary-thyroid system. All thyroxine (T_4) and some triiodothyronine (T_3) are produced by the thyroid gland, and their production there is stimulated by thyroid-stimulating hormone (TSH, thyrotropin), a product of the anterior pituitary gland. Some T_4 is converted to T_3 in other tissues, including the pituitary gland and the hypothalamus. T_3 inhibits pituitary secretion of TSH, and hypothalamic secretion of thyrotropin-releasing hormone (TRH), which stimulates TSH secretion. The interplay between T_3 and TSH maintains thyroid hormone production within a narrow range. (+), stimulation; (−), inhibition. Sources: Liver illustration reprinted by permission of Media Lab, University of Wisconsin-Madison (2004), copyright 2003; thyroid illustration in Yale New Haven Health (2004) reprinted by permission of Nucleus Medical Art, copyright 2005. All rights reserved. www.nucleusinc.com.

hypothalamus are inhibited by T_4 and T_3 and stimulated in their absence. People with thyrotropin-releasing hormone deficiency have hypothyroidism because they have TSH deficiency.

Thyrotropin

TSH (thyrotropin) is synthesized and secreted by specific cells (thyrotroph cells) of the anterior pituitary gland (Figure 2-4). It stimulates virtually every step of T_4 and T_3 synthesis and secretion by the thyroid gland, including NIS activity, synthesis of thyroglobulin, formation of T_4 and T_3 in thyroglobulin, and breakdown of thyroglobulin and release of T_4 and T_3 into the bloodstream. It also stimulates the blood supply and growth of the thyroid gland.

The secretion of TSH is inhibited by small increases in serum T_4 and T_3 concentrations, and it increases in response to small decreases in serum T_4 and T_3 concentrations. This tight control of TSH secretion results in maintenance of T_4 and T_3 production and secretion by the thyroid gland within very narrow limits. Thyroid gland function decreases in people with TSH deficiency, as a result of a pituitary gland disorder, just as it does in people with thyroid disease.

Regulation of Extrathyroid Production of Triiodothyronine

The tissue distribution and regulation of the deiodinases that catalyze the conversion of T_4 to T_3, the conversion of T_4 to reverse T_3, and the conversion of T_3 to 3,3'-diiodothyronine differ (Bianco et al. 2002). The activity of type 1 deiodinase, the predominant deiodinating enzyme in the liver and kidney, is decreased by thyroid hormone deficiency (hypothyroidism) and increased by thyroid hormone excess (hyperthyroidism). The activity of type 2 deiodinase—the predominant deiodinating enzyme in the brain, pituitary, and muscle—is increased by thyroid hormone deficiency and decreased by thyroid hormone excess. The activity of those deiodinases is also altered by nonthyroid illness, caloric deprivation, other hormones, and drugs, and it differs between fetuses and adults.

IODIDE NUTRITION AND METABOLISM

Iodide is essential for the production of T_4 and T_3 by the thyroid gland. It can be obtained only by ingestion of foods that naturally contain it or of foods to which iodide is added during processing (iodization). Foods rich in iodide include seafood and sea products (kelp and seaweed), dairy products, eggs, commercial bakery products, and some vegetables. Sea salt also contains iodide, and iodized salt is widely available (and mandated by law in many countries).

Dietary iodide is absorbed and distributed rapidly (as iodide) in the extracellular fluid, which also contains iodide released from the thyroid gland during hormone secretion and from extrathyroid deiodination of T_4 and T_3. T_4 and T_3 are eventually completely deiodinated, and the iodide returned to the circulation. Iodide leaves the circulation by transport into the thyroid or by excretion in the urine. In healthy adults in the United States, serum inorganic iodide concentrations range from 1 to 2 μg per deciliter (dL).

The World Health Organization (WHO) recommends a dietary intake of 150 μg per day for adults, 200 μg per day for pregnant women, 90-120 μg per day for children 2-11 years old, and 50 μg per day for infants less than 2 years old (WHO 1996). The Institute of Medicine of the National Academies recommends a slightly higher intake, 220 μg per day, for pregnant women (IOM 2000).

At those intakes there are no clinical or biochemical manifestations of thyroid dysfunction. Lower intakes are associated with increasing frequency of thyroid enlargement (goiter), biochemical evidence of thyroid hormone deficiency, and ultimately, in people with severe iodide deficiency, hypothyroidism. WHO considers people whose iodide intake ranges from 50 to 99 μg per day to have mild iodide deficiency, those whose intake ranges from 20 to 49 μg per day to have moderate iodide deficiency, and those whose intake is less than 20 μg per day to have severe iodide deficiency. (In studies of iodide nutrition, iodide intake is not measured directly but is usually estimated as the amount of iodide in a liter of urine, because excretion of iodide by the kidneys is by far the most important route of loss of iodide from the body.)

From 1988 to 1994, iodide intake in the United States averaged about 150 μg per day on the basis of single spot (untimed) measurements of urinary iodide excretion in 20,369 people 6-74 years old (median urinary iodide excretion, 145 μg/L) (Hollowell et al. 1998); the median value was about 50% lower than the value in 1971-1974. The value was less than 50 μg per day in 12% of adults (15% of women of childbearing age and 7% of pregnant women). It should be noted that the distribution of iodide values measured in spot urine samples is broader than values measured repeatedly in individual subjects (Andersen et al. 2001); this leads to overestimation of the number of subjects with both low and high values. Furthermore, among people 15-44 years old, including pregnant women, there were no differences in serum TSH and T_4 concentrations between those with urinary iodide values less than 50 μg/L and those with higher values (Soldin et al. 2005); apparently, the iodide intake in the former group was not low enough to cause a fall in T_4 secretion.

The reasons for the decrease in iodide intake in the United States between 1971-1974 and 1988-1994 are not known precisely, but they include lower salt intake (iodized salt contains iodide at 76 μg/g), less use of iodide in the baking and dairy industries, and a decrease in the addition of iodide to animal feed. More recent data suggest that the decline has ceased; in a national survey in 2000, the median value was 161 μg/L as compared with 145 μg/L in 1988-1994 (NCHS 2002).

Iodide deficiency is more prevalent in many other countries but can be largely ameliorated by salt iodization. So far as the committee is aware, there are no reports of perchlorate exposure in areas of iodide deficiency.

PERTURBATIONS OF THYROID HORMONE PRODUCTION

Severe iodide deficiency is one of many conditions that can reduce T_4 and T_3 production (and is the most common worldwide). Others include iodide excess, various drugs, congenital abnormalities of development of the thyroid gland, congenital deficiencies of the components of the T_4 and T_3 synthesis pathway, and many diseases that damage the thyroid gland. The range of thyroid hormone deficiency that occurs in those conditions varies greatly, from almost clinically undetectable and fully compensated by the mechanism described below to severe hypothyroidism. Among infants, hypothyroidism can result in severe abnormalities in neural and skeletal development; in adults, it can result in substantial disability and rarely hypothyroid coma.

When thyroid gland synthesis and secretion of T_4 and T_3 fall as a result of iodide deficiency or any other cause, the serum concentrations of the hormones fall. That results in a prompt increase in TSH secretion. If the thyroid is severely damaged or has been removed surgically or if the dose of an offending drug is high enough to block synthesis and secretion of the hormones completely, TSH has little effect. Serum T_4 and T_3 concentrations continue to fall, and although TSH secretion increases further, severe hypothyroidism occurs. If the problem is iodide deficiency or if thyroid damage or drug blockade of T_4 and T_3 synthesis and secretion is incomplete, the initial increase in serum TSH concentrations stimulates synthesis and secretion of the two thyroid hormones enough to raise their serum concentrations to normal or near normal. The rise in turn lowers TSH secretion to, or almost to, its original level. The person has few, if any, symptoms or signs of hypothyroidism, although the thyroid gland may enlarge. Indeed, thyroid enlargement may be the only evidence that T_4 and T_3 production was low and TSH secretion was high at an earlier time.

In summary, there is a potent mechanism—increased TSH secretion by the pituitary gland—to compensate for thyroid hormone deficiency. The mechanism is activated by small decreases in T_4 and T_3 production. It effectively restores thyroid hormone production to normal or near normal even when the initial insult is substantial, for example, a large fall in iodide intake (see next section).

There is another compensatory mechanism—the effect of hypothyroidism to increase conversion of T_4 to the more active T_3 in some tissues, especially the brain. The role of that intracellular compensatory mechanism is hard to measure, but it almost certainly contributes to the apparent normality of people, including infants, with decreased T_4 production.

Iodide Deficiency and Other Perturbations

The varied effects of iodide deficiency, depending on its severity, provide an example of the compensatory mechanism. People with mild iodide deficiency, as defined above, have normal serum T_4 and TSH concentrations, but about 5-10% have some thyroid enlargement. People with moderate iodide deficiency also have normal serum T_4 and TSH concentrations, but about 20-30% have thyroid enlargement. And people with severe iodide deficiency may have slightly low serum T_4 concentrations and high serum TSH concentrations, and over 30% have thyroid enlargement; overt hypothyroidism occurs only if iodide intake is below about 5-10 μg per day (Delange 2000). As iodide intake declines, thyroid uptake of iodide increases because of an increase in the number of NIS molecules. Those changes are intrinsic parts of the compensatory response; they are facilitated by an increase in TSH secretion but probably can occur in the absence of an increase. In addition, there is a shift to production of T_3, which contains less iodide but is more active than T_4. In summary, there is remarkable compensation for the effects of iodide deficiency so that even in the presence of very low iodide intake normal or near-normal thyroid hormone production and TSH production are maintained. However, severe deficiency in iodide intake (below 20 μg per day) in pregnant women may result in major neurodevelopmental deficits and goiter in their offspring. Similar iodide deficiency in infants and children may result in smaller but still important neurodevelopmental deficits (Delange 2000).

Iodide excess and therapy with lithium carbonate provide additional examples of the compensatory mechanism. When ingested at 1 mg per day (1,000 μg per day) or more, iodide has an antithyroid action in healthy subjects. In 1-2 weeks, serum T_4 and T_3 concentrations fall by 10-15% and serum TSH concentrations increase by about 50% (Roti and Vagenakis 2000). Those changes persist, but do not worsen, if intake of excess iodide is continued. A similar pattern of changes in serum T_4, T_3, and TSH concentrations occurs in people with psychiatric disease who are treated with lithium carbonate, a drug with antithyroid actions. The same pattern of changes in serum T_4, T_3, and TSH concentrations may be found in people with many thyroid disorders, including autoimmune thyroiditis, in which the thyroid is damaged by immune mechanisms; tumors of the head and neck regions that have been treated with radiation; hyperthyroidism that has been treated with radioactive iodide; and surgical removal of one side of the thyroid gland. In summary, many substances and conditions lower T_4 and T_3 secretion and result in a rise in TSH secretion. If the thyroid gland is not seriously damaged, serum T_4 and T_3 concentrations may return to normal, or so near to normal that there are few or no consequences.

Clinical Consequences of Perturbations of Thyroid Hormone Production

Compensation for iodide deficiency or other perturbations in thyroid hormone production, as described above, is the rule. In those cases, adults have no clinical consequences; they have normal serum T_4 and TSH concentrations and no clinical abnormalities. If the perturbation is greater, hypothyroidism occurs.

Multiple terms are used to describe hypothyroidism, according to its severity, causes, duration, and the affected persons. They include the following:

- *Subclinical hypothyroidism* is defined as high serum TSH concentration and normal serum T_4 concentration.
- *Overt hypothyroidism* is defined as high serum TSH concentration and low serum T_4 concentration.
- *Primary hypothyroidism* is hypothyroidism resulting from thyroid disease or a condition (iodide deficiency) or drug that decreases thyroid hormone production.
- *Central (or pituitary) hypothyroidism* is hypothyroidism resulting from deficiency of TSH or thyrotropin-releasing hormone.

- Hypothyroidism may be permanent or transient—lasting a few weeks or months or until reversal of the causative condition, for example, by treatment of iodide deficiency or disappearance of antibodies that block the action of TSH.
- Hypothyroidism may be congenital (present at birth) or acquired later in life.

In the United States, subclinical hypothyroidism is found in 4-8.5% of adults (Surks et al. 2004) and 2.5% of pregnant women (Klein et al. 1991); the frequency is higher in parts of the world that have severe iodide deficiency. People with subclinical hypothyroidism have few or no symptoms of hypothyroidism (Surks et al. 2004). They may have mild hypercholesterolemia and small decreases in cardiovascular function (Biondi et al. 2002), but overall cardiovascular-disease morbidity and mortality are not increased (Vanderpump et al. 1996; Parle et al. 2001). However, people with subclinical hypothyroidism caused by autoimmune thyroiditis have progression to overt hypothyroidism at a rate of 2-4% per year (Vanderpump et al. 1995). There is general agreement among endocrinologists that women with subclinical hypothyroidism who are pregnant or planning pregnancy should be treated with T_4 but little agreement regarding treatment of others with the disorder.

Thyroid hormone production must fall substantially and, more important, must remain low for a prolonged period for adverse effects to occur (see section Maternal Hypothyroidism for discussion of effects in pregnant women). Removal of half a person's thyroid is not associated with symptoms, because the resulting thyroid deficiency is not prolonged (Matte et al. 1981). The minimum extent and duration of the fall in thyroid hormone production that has adverse effects are not known. It is known that thyroid hormone production can fall enough to increase TSH secretion yet remain within the normal range (subclinical hypothyroidism, as defined above). Some of the people in whom that happens have slightly high serum TSH concentrations (4.6 to10 milliunits [mU] per liter; normal range, 0.5 to 4.5) but no symptoms or other manifestations of hypothyroidism (Surks et al. 2004). Among people with overt hypothyroidism, there is a moderate inverse correlation between serum T_4 concentration and severity of symptoms, increase in serum TSH concentration, and degree of abnormality of several physiologic and biochemical measures of thyroid hormone action (for example, reflex time and serum cholesterol concentration) (Staub et al. 1992). In studies of people being treated for hypothyroidism, changes in the daily dose of T_4 of 25 μg (about one quarter of daily production in healthy adults) have small effects not only on TSH secretion but also on well-being (Carr et al. 1988) and resting metabolic rate (Al-Adsani et al. 1997).

Factors Affecting Susceptibility to Substances with Antithyroid Activity

People who have a condition that reduces T_4 and T_3 production by the thyroid would be expected to be more sensitive to any additional factor that might further reduce production of the two hormones. The additional factor might be another thyroid disorder or treatment with a drug or other substance that has antithyroid activity. Autoimmune thyroiditis is the most common thyroid disorder in the United States, and people who have it are known to be more vulnerable to the antithyroid activity of iodide or lithium carbonate (Roti and Vagenakis 2000). People who have iodide deficiency would also be expected to be more susceptible to antithyroid substances. For example, in some regions of the world (notably Africa) where iodide intake is low, a high dietary intake of cassava, which contains thiocyanate, accentuates the iodide deficiency (Delange 2000). Perchlorate, if present in sufficient concentrations, might have the same effect, although an interaction between iodide deficiency and perchlorate has not been documented.

Thyroid Overactivity

The converse sequence of changes occurs when serum T_4 and T_3 concentrations rise, whether because of thyroid inflammation and release of thyroid hormones stored in the thyroid gland; because of a thyroid disorder, such as a nodular goiter, that results in secretion of thyroid hormones independently of TSH; because of Graves disease, in which the thyroid gland is stimulated by antibodies that act like TSH; or because of ingestion of T_4 or T_3 tablets. When serum T_4 and T_3 concentrations rise, TSH secretion decreases. T_4 and T_3 production by any normal thyroid tissue then falls. Thus, the person is to some extent protected against hyperthyroidism by the normal negative feedback mechanism.

HYPOTHALAMIC-PITUITARY-THYROID DEVELOPMENT AND FUNCTION IN FETUSES AND INFANTS

Development of the Hypothalamic-Pituitary-Thyroid System in Fetuses

Before the onset of human fetal thyroid function at 10-12 weeks of gestation, T_4, presumably of maternal origin, can be detected in the coelomic fluid of 6-week-old embryos (Calvo et al. 2002). It can also be detected in the brains of 10-week-old fetuses, as can deiodinases and T_3 receptors (Kilby et al. 2000). That maternally derived T_4 is probably essential for normal early development.

The thyroid gland, pituitary, and hypothalamus form during the first trimester of gestation (Fisher and Brown 2000). The thyroid gland is derived from cells that originate in the floor of the primitive pharynx and then descend into the neck, where they divide and move laterally, forming the two lobes of the thyroid. Formation of the precursors of the thyroid follicular cells and the growth and descent of these precursor cells into the neck result from the coordinate action of multiple transcription factors (molecules that regulate gene function), hormones, and growth factors. Primitive thyroid follicles, iodide uptake, and thyroglobulin are first detected at 10-12 weeks of fetal life. Low concentrations of T_4 and T_3 can be detected in fetal serum soon thereafter.

The iodide needed for fetal thyroid hormone synthesis must be provided by the mother. Placental tissue contains a small amount of the NIS, but its importance in maternal-to-fetal transfer of iodide is probably small. That women who have mutations in the NIS can have normal children suggests that absence of transporters in tissue of maternal origin does not limit the transfer of iodide from mother to fetus (Kosugi et al. 2002).

The anterior pituitary gland, site of production of TSH, develops from the floor of the primitive forebrain and an outgrowth of the primitive oral cavity (Rathke's pouch). Although the structures are visible by 5 weeks, a morphologically mature pituitary gland and its vascular connections with the hypothalamus above it are not identifiable until 14 weeks. Low concentrations of TSH can be detected in fetal serum from 12 weeks to about 18-20 weeks, after which they rise substantially. The hypothalamic centers that produce thyrotropin-releasing hormone and the hormone itself are not identifiable until 15-18 weeks.

Maturation of the Hypothalamic-Pituitary-Thyroid Axis

At 18-20 weeks, fetal serum TSH concentration begins to rise, and then serum total and free T_4 concentrations rise progressively (Fisher and Brown 2000). The progressive increase in serum total T_4 concentration is due to a concomitant rise in serum TBG concentration, caused by an increase in production of the protein by the fetal liver and by increased thyroid production of T_4. The increase in TSH secretion is caused by rising thyrotropin-releasing hormone secretion from the maturing hypothalamus and is required to stimulate the fetal thyroid gland to produce the increasing amounts of T_4 needed to sustain the rise in serum total and free T_4 concentrations.

The progressive maturation of thyrotropin-releasing hormone and TSH secretion during the latter part of gestation and early postnatal life results in an increasing capacity to respond to both decreases and increases in serum T_4 concentrations. Thus, as in adults, TSH secretion in fetuses increases if fetal thyroid secretion is inhibited as a result of a thyroid disorder or iodide deficiency or the transfer of a substance that has antithyroid activity. Conversely, fetal TSH secretion is inhibited if serum T_4 concentration increases.

Fetal T_3 and reverse T_3 metabolism is different from that in adults. Fetal serum T_3 concentrations are low until about 32 weeks and thereafter rise slowly until birth. Conversely, serum reverse T_3 concentrations are high and decline near birth. Fetal serum T_3 concentrations are low because fetal tissues and the placenta have low type 1 deiodinase activity (which converts T_4 to T_3) and high type 3 deiodinase activity (which converts T_3 to diiodothyronine) (Bianco et al. 2002).

The fetal hypothalamic-pituitary-thyroid system does not function completely independently of that of the mother. Thyrotropin-releasing hormone and TSH do not cross the placenta in appreciable amounts, but T_4 and T_3 do cross the placenta, as noted above. In infants who cannot synthesize any T_4 or T_3, serum T_4 concentrations at birth are 20-40% of those of healthy infants (Vulsma et al. 1989). Maternal T_4 and T_3 undoubtedly contribute to the serum concentrations of the two hormones in healthy fetuses but not as much as in fetuses that are unable to produce the hormones themselves.

Development of Tissue Deiodinases and Triiodothyronine-Nuclear Receptors

The tissue distribution and activity of the deiodinase enzymes in fetuses differ from those in adults. Studies in humans are very limited but show that type 3 deiodinase, which inactivates T_4 and T_3, can be detected in fetal liver and brain in 10-week-old fetuses. Its content in the liver decreases progressively thereafter during fetal life, and it decreases further in the first weeks after birth (Fisher and Brown 2000). Type 1 deiodinase activity has been detected in the liver of 20-week-old fetuses, and its activity gradually increases thereafter, not only during gestation but also after birth. The high fetal type 3 deiodinase activity, with high placental type 3 deiodinase activity, explains why fetal serum T_3 concentrations are low early in gestation, increase only slowly until birth, and then increase rapidly (and why fetal serum reverse T_3 concentrations change in the opposite way) (Bianco et al. 2002). Type 2 deiodinase is present in fetal tissues, and its content, at least in animals, increases during gestation and after birth. Indeed, in animals, the patterns of deiodinase activity differ in different tissues, even in different regions of tissues, such as the brain, at different times during development. Those findings indicate that there are variations in local production of T_3 at different times during normal development and that correct timing of the changes is critical for normal development.

The tissue distribution and timing of appearance of T_3-nuclear receptors in fetuses also differs from those in adults. T_3 receptors can be detected in embryonic animal stem cells; in the brains of 10- to 11-week-old human fetuses, probably before the fetal thyroid begins to secrete T_4 (Kilby et al. 2000); and in the liver, heart, and lungs of 16- to 18-week-old fetuses.

Thyroid Function in Infants

Term Infants

At birth, there is a dramatic increase in TSH secretion in healthy term infants, a consequence of the relative hypothermia of the extrauterine environment. Serum TSH concentrations rise abruptly, from 3 to 8 mU/L at delivery to 50 to 70 mU/L in 30-60 minutes (min) (Fisher and Brown 2000). The concentrations then fall rapidly for 24 hr, and more slowly thereafter, to less than about 10-12 mU/L at the end of the first week of life. The thyroid stimulation that results from the surge in TSH secretion raises serum T_4 concentrations by about 50% in 24 hr, after which they gradually decline. Serum T_3 concentrations increase by 300-400% in 24 hr, not only because of the increase in TSH secretion but also because of an increase in type 1 deiodinase activity and a decrease in type 3 deiodinase activity in the infants' tissues. Serum T_4 and T_3 concentrations gradually fall in the first 4 weeks of life to values that are slightly higher than those in adults.

Preterm Infants

At birth, the functioning of the hypothalamic-pituitary-thyroid system is slightly lower in preterm infants than in term infants, but for the most part it is appropriate for their gestational age. Their serum free T_4 concentrations are slightly lower than those of fetuses of comparable age in utero. After delivery, serum TSH and T_4 concentrations increase, as in infants born at term, but the increases are slightly smaller. In very premature infants (less than 32 weeks of gestation), there is no surge of TSH secretion after birth, and serum T_4 concentrations fall during the first week of life, the degree of decline reflecting the degree of prematurity (Oden and Freemark 2002). Even more premature infants (less than 30 weeks) may have very low serum T_4 concentrations at birth and no postnatal rise. Their serum total T_4 concentrations tend to be more abnormal than their serum free T_4 concentrations. The difference is indicative of decreased production of TBG or other binding proteins as a consequence of immature liver function. In infants who are healthy and grow well, serum T_4 concentrations gradually increase to values similar to or higher than those at birth in 4-6 weeks.

In addition to the aforementioned changes in serum TSH and T_4 concentrations, preterm infants have low serum T_3 concentrations. That is due to immaturity of type 1 deiodinase, inhibition of the activity of this enzyme by nonthyroid illness (for example, the respiratory distress syndrome), or both.

The causes for the postnatal fall in serum T_4 concentrations in very premature infants include clearance of maternal T_4, decreased serum binding of T_4, and immaturity of the hypothalamic-pituitary-thyroid system. In those infants, TSH secretion does not increase as much in response to low serum T_4 concentrations as it does in more mature infants. Furthermore, preterm infants are more likely to suffer from systemic illness, such as the respiratory distress syndrome, and to be treated with medications, such as glucocorticoids, that reduce TSH or thyroid hormones. All those situations tend to reduce serum TSH and T_4 concentrations. In addition, preterm infants are more sensitive to the effects of iodide deficiency and iodide excess, and the incidence of transient hypothyroidism with high serum TSH is increased (Frank et al. 1996). In view of the critical dependence of brain maturation on thyroid hormone at this age, any period of low serum free T_4 concentrations could impair neurologic or neuropsychologic development. However, in most preterm infants, the hormonal abnormalities are transient, and the infants do not benefit from T_4 therapy—as assessed by mental development, neurologic function, or growth—compared with placebo (van Wassenaer et al. 1997; Oden and Freemark 2002).

Thyroid Hormone Actions in Developing Fetuses and Newborn Infants

T_3 is required for normal development of the central nervous system. Its actions include stimulation of the development and growth of neurons (nerve cells) and supporting (glial) cells, the formation of connections (synapses) between neurons, the formation of the myelin sheaths that surround neuronal processes, and the development of the neurotransmitters that transmit signals from one nerve cell to another (Fisher and Brown 2000). T_3 stimulates the transcription of several genes whose products are important for neural development, including the genes for myelin basic protein, a cerebellar Purkinje-cell protein, a calcium-binding protein, and neurogranin (Oppenheimer and Schwartz 1997). In hypothyroid animals, the expression of those genes is delayed, but most are ultimately expressed to the same extent as in normal animals. The resulting abnormalities in neurologic and neuropsychologic development, although variable and determined at least in part by when the deficiency occurred, are permanent; this indicates that the correct timing of the expression of those and other genes in the brain during development is critical. However, the linkage between the biochemical abnormalities in the brain and the developmental abnormalities is far from clear.

T_4 and T_3 also are required for normal skeletal development and growth. Bone cells have T_3 receptors, and T_3 stimulates bone formation and the appearance of the epiphyseal centers that are needed for normal growth of long bones. T_3 also stimulates the production of pituitary growth hormone and insulin-like growth factor. Treatment with T_4 leads to resumption of bone growth and skeletal maturation, but severely affected infants are unlikely to have normal stature.

Effects of Perturbations of Maternal, Fetal, and Child Thyroid Function on Fetal and Child Development

The clinical manifestations of hypothyroidism in infants vary widely, according to whether the mother, the fetus, or both have hypothyroidism and how long it persists after birth. The abnormalities are greatest when both mother and fetus are affected; this is most likely to occur in regions of severe iodide deficiency. The consequences of severe combined maternal and fetal hypothyroidism during fetal life and in newborn infants include microcephaly (small brain), mental retardation, deaf-mutism, paraplegia or quadriplegia, and movement disorders. Those abnormalities are not reversible by treatment with T_4 (Foley 2000). However, the abnormalities can be largely prevented by administration of iodide to the mothers before or during the first trimester and early part of the second trimester of pregnancy (Pharoah 1993; Cao et al. 1994). That finding underlies the importance of the availability of T_4 from the mother before fetal thyroid secretion begins, as noted above.

The infants of mothers who have mild iodide deficiency have larger thyroid glands and higher serum TSH or thyroglobulin concentrations at birth than do infants of mothers whose iodide intake is higher (Glinoer et al. 1995; Kung et al. 2000). Otherwise, they appear to be neurologically and physically normal.

Newborn infants who have hypothyroidism may have other abnormalities, including lethargy, poor muscle tone, poor feeding, constipation, and persistent jaundice, if not at birth then thereafter. The changes are similar to those which occur in older children and adults who have hypothyroidism, and, in contrast with the neurologic abnormalities, they are reversible with adequate T_4 treatment.

Maternal Hypothyroidism

Pregnant women who have overt hypothyroidism and are adequately treated have normal pregnancies, and their infants develop normally (Liu et al. 1994). Pregnant women who have subclinical hypothyroidism or overt hypothyroidism and are inadequately treated or not treated at all have an increased risk of fetal loss (Allan et al. 2000; Abalovich et al. 2002). The infants of those mothers who do not miscarry have normal thyroid function at birth and thereafter, but their neurodevelopment may be slightly impaired. A prospective study of seven infants born to mothers who had subclinical hypothyroidism during pregnancy and six infants born to mothers who had normal thyroid function found that the former had lower scores on the Bayley Mental Developmental Index at the ages of 6 and 12 months but not at 24 months; the scores on the Bayley Psychomotor Development Index were similar at all three times (Smit et al. 2000). Another study compared 7- to 9-year-old children born to 62 women who had subclinical hypothyroidism during the second trimester of pregnancy with 124 children born to women who had normal thyroid function (Haddow et al. 1999). The mean full-scale IQ score was 4 points lower in the former group, and 15% had scores of 85 or lower, compared with 5% of the control children.

The infants of mothers who have low serum free T_4 concentrations early in pregnancy also may have slightly impaired neurodevelopment. Among

220 infants tested with the Bayley Scales of Infant Development at the age of 10 months, the 22 infants whose mothers had serum free T_4 concentrations in the lowest 10th percentile (but normal serum TSH concentrations) at 12 weeks of gestation scored lower on the Psychomotor Development Index (by 7 points, 93% vs 100%) but not the Mental Developmental Index (Pop et al. 1999). In a second study, 57 infants whose mothers had serum free T_4 values in the lowest 10th percentile and 58 infants whose mothers had serum free T_4 values in the 50th to 90th percentile were tested in the same way at the ages of 1 and 2 years. The scores on both indexes were slightly but statistically significantly lower (mean differences ranged from 4 to 6 points of 100 points) at both times in the infants of the mothers who had low serum free T_4 values (Pop et al. 2003).

Those studies, although not definitive, suggest an effect on development in infants whose mothers had subclinical hypothyroidism or low-normal serum free T_4 concentrations during pregnancy, but they have limitations. The differences in test scores were small, and the scores could be confounded by socioeconomic, educational, and other differences between the study groups. Moreover, the results contrast with the normal development of the infants of mothers who had overt hypothyroidism (Liu et al. 1994). Nonetheless, if confirmed, they emphasize the potential vulnerability of fetuses to decreases in maternal thyroid function.

Fetal and Neonatal Hypothyroidism

Infants who have congenital hypothyroidism usually appear normal at birth (Foley 2000). Their serum T_4 concentrations are low, but not very low and indicate that some maternal T_4 crossed the placenta. Their serum TSH concentrations are high and rise further soon after birth as the maternally derived T_4 is metabolized and its concentration in the infants' serum falls. Those infants can be identified as having hypothyroidism if screened by measurements of TSH or T_4 in blood collected 24-96 hr after birth; such screening has been in place in the United States for about 25 years. Infants so identified by neonatal screening have normal neural development and growth if aggressive T_4 treatment is started within the first 2 or 3 weeks after birth.

After birth, not only maternal T_4 and T_3 but also other maternal factors that might have affected fetal thyroid secretion are cleared from the infant's circulation. Whether those substances alter a newborn infant's thyroid function depends on the dose and rate of clearance of the substance and the prematurity of the infant. The efficacy of prompt T_4 treatment of newborn infants found by screening to have hypothyroidism makes it highly unlikely that any rapidly cleared substance that reached the fetus from the mother and reduced thyroid secretion in the fetus in utero but no longer reached the infant after birth could cause postnatal hypothyroidism of sufficient severity to cause permanent developmental delay. That conclusion is born out by the uncommon clinical situation described below.

Hyperthyroidism occurs in about one in 2,000 pregnant women. Some of the affected women require treatment with an antithyroid drug throughout their pregnancies. The antithyroid drugs propylthiouracil and methimazole may cross the placenta in sufficient quantities to cause transient fetal hypothyroidism. After birth, no more drug reaches the infants, the hormonal changes that can be detected in utero disappear rapidly, and the infants develop normally (Eisenstein et al. 1992). Years ago, some pregnant women who had hyperthyroidism were treated successfully with potassium perchlorate. Most of the infants were normal, but one had slight thyroid enlargement that disappeared soon after birth. No other abnormalities at birth were reported, and the infants were not followed thereafter (Crooks and Wayne 1960).

Iodide Nutrition in Childhood

Adequate iodide intake during infancy and childhood is also important; children with moderate iodide deficiency have learning disabilities and do less well on tests of mental and psychomotor performance than do children with adequate iodide intake (Tiwari et al. 1996; van den Briel et al. 2000; Santiago-Fernandez et al. 2004). In the most extensive study of this topic, of 1,221 Spanish children (mean age, 10.8 years), 30% of children with urinary iodide excretion less than 25 μg/L had an IQ at or below the 25th percentile, compared with 16% of children with urinary iodide excretion of more than 150 μg/L (Santiago-Fernandez et al. 2004). However, urinary iodide values were positively correlated with serum TSH values and not correlated with serum free T_4 and free T_3 values, and the results were not adjusted for parental socioeconomic or educational status.

PERCHLORATE AND THE THYROID

Perchlorate potentially can affect thyroid function because of its ability to block the transport of iodide into thyroid follicular cells. As noted above, it does so by competitively inhibiting iodide transport by the NIS in the plasma membrane of these cells. The fact that the inhibition is competitive means that it can be overcome by higher concentrations of iodide, and, in laboratory studies, perchlorate did not inhibit uptake of iodide by thyroid tissue when high concentrations of iodide were present (Wolff 1998).

After recognition in the 1950s of the ability of perchlorate to block uptake of iodide in animal and then human thyroid tissue (Stanbury and Wyngaarden 1952), it was given on a long-term basis to patients who had hyperthyroidism, with the goal of reducing synthesis and secretion of T_4 and T_3. The effects of perchlorate therapy in hyperthyroid patients and of perchlorate given prospectively to healthy subjects are reviewed in the following sections.

Therapeutic Uses of Perchlorate

The medical literature of the 1960s contains reports of successful treatment with potassium perchlorate of more than 1,000 patients who had hyperthyroidism caused by Graves disease or nodular goiter. The potassium perchlorate was given in doses of 400-2,000 mg per day for many weeks or months (Crooks and Wayne 1960; Morgans and Trotter 1960; summarized in Wolff 1998 and Soldin et al. 2001). One patient was treated for 22 years at 200 mg per day (Connell 1981). The patients included 12 pregnant women who were treated with 600-1,000 mg per day. The only adverse effect was slight thyroid enlargement in one infant, which decreased soon after birth, as noted above (Crooks and Wayne 1960).

Potassium perchlorate treatment of hyperthyroid patients was for the most part safe, although some had nausea and vomiting (gastric inflammation), skin rashes, fever, lymph node enlargement, and kidney dysfunction. The frequency of those side effects was dose-dependent; they occurred in 3-4% of patients taking 400-600 mg per day, and in 16-18% of patients taking 1,000-2,000 mg per day (Crooks and Wayne 1960; Morgans and Trotter 1960). Thirteen patients who had taken 400-1,000 mg per day for 2-20 weeks developed aplastic anemia or agranulocytosis (cessation of production of red blood cells or white blood cells, respectively), and seven of them died (summarized in Soldin et al. 2001). As a result of the latter events, and perhaps also because of the contemporaneous development of other antithyroid drugs, the use of perchlorate to treat patients for hyperthyroidism largely ceased by the middle to late 1960s.

The most recent report of long-term potassium perchlorate treatment of patients for hyperthyroidism caused by Graves disease was in 1984. In that study, 18 patients were treated initially with 900 mg per day (Wenzel and Lente 1984). As serum thyroid hormone concentrations declined, the dose of potassium perchlorate was reduced to an average of 93 mg per day at 12 months. Thereafter, the patients received 40-120 mg per day (perchlorate at 0.41-1.2 mg/kg of body weight per day for a 70-kg person) for 12 months. During that 12-month period, all the patients had normal serum T_4 and T_3 concentrations, and the majority had normal serum concentrations of TSH-receptor stimulating antibodies, the cause of hyperthyroidism in patients who had Graves disease; this indicated that they no longer had Graves disease. There is no mention of side effects in the report, but all the patients treated with perchlorate completed the study. Given that most of the patients did not have high serum concentrations of TSH-receptor stimulating antibodies during the second year of perchlorate therapy, the results strongly suggest that moderate doses of perchlorate given chronically do not cause hypothyroidism.

Potassium perchlorate treatment has been revived in recent years as a treatment for patients who have a type of hyperthyroidism caused by amiodarone, an iodinated drug used to treat patients for abnormal cardiac rhythms. Amiodarone can cause two types of hyperthyroidism: one from the excess iodide and the other from thyroid inflammation. The iodide-induced type of hyperthyroidism usually occurs in patients who have a pre-existing nodular goiter, and it has proved difficult to treat with standard antithyroid drugs. Cessation of amiodarone is not helpful, because the drug is not completely excreted for many months. In affected patients, perchlorate therapy can be helpful because of its ability to prevent iodide uptake by thyroid tissue. Potassium perchlorate at 200-1,000 mg per day has been given for several weeks or months. There have been no reports of adverse effects, and in some of the studies blood counts and kidney function were monitored and remained normal (Martino et al. 1986; Newnham et al. 1988; Reichert and de Rooy 1989; Bartalena et al. 1996). Most patients were treated simultaneously with a standard antithyroid drug (methimazole or propylthiouracil), so the benefit cannot be ascribed to perchlorate alone.

Perchlorate has also been given prophylactically to patients who have multinodular goiter to prevent iodide-induced hyperthyroidism. They were patients who were to receive an iodinated contrast agent, which, like amiodarone, contains large amounts of iodide, to visualize blood vessels as part of radiographic or computed-tomography studies. In a study in which 51 patients who had multinodular goiter were randomly assigned to receive no treatment or methimazole or sodium perchlorate (900 mg per day) for 14 days starting on the day of a radiographic study, two patients in the control group became hyperthyroid, compared with one patient in each treatment group (Nolte et al. 1996). There were no side effects of the sodium perchlorate.

There are no reports of the appearance of a new thyroid disorder, thyroid nodules, or thyroid carcinoma in any patient treated with potassium perchlorate for hyperthyroidism. Iodide deficiency in the thyroid gland, a possible consequence of perchlorate administration or exposure, is not associated with an increase in thyroid cancer (Schlumberger 1998). In hyperthyroid patients treated with antithyroid drugs, there was no increase in thyroid cancer mortality (Ron et al. 1998).

Clinical Pharmacology and Effects of Perchlorate in Healthy Humans

Perchlorate is usually administered as potassium perchlorate. It is rapidly absorbed after ingestion, and peak serum perchlorate concentrations are reached in 3 hr. Its half-life in serum is about 6-8 hr, and it is rapidly eliminated unchanged from the body, primarily in urine.

Five studies have been conducted in which perchlorate was given to healthy subjects for various periods, and its effects on thyroid function were determined. In one study, five healthy men were given 200 μg of iodide daily for 4 weeks and then 900 mg of potassium perchlorate daily (perchlorate at about 9 mg/kg per day for a 70-kg person) for 4 weeks. At the end of each 4-week period, there were no differences in serum T_4 and T_3 concentrations or thyroid volume. The mean 24-hr serum TSH concentrations (1.0 vs 1.8 mU/L), serum free T_4 concentrations (14.3 vs 15.7 picomoles/L), and total thyroid iodide content (3.0 vs 4.0 millimoles/mL) were slightly lower at the end of the perchlorate period than at the end of the iodide period (Brabant et al. 1992). Note that the serum TSH values were lower at the end of the perchlorate period, indicating lack of an antithyroid effect.

In the second study (Lawrence et al. 2000), nine healthy men 22-30 years old ingested 10 mg of potassium perchlorate (perchlorate at about 0.10 mg/kg per day for a 70-kg person) in 1 L of water daily for 14 days. Their serum perchlorate concentrations averaged 0.6 μg/mL, and the average urinary perchlorate excretion was 7.7 mg per day. Urinary iodide values did not change (the mean baseline value was 254 μg per day). There were no changes in serum T_4, T_3, or TSH concentrations during the 14-day period of perchlorate ingestion. The 24-hr thyroid uptake of radioactive iodide (iodide-123) was measured three times. At baseline, the mean value was 23.6% of the administered dose; it decreased to 14.0% of the dose after perchlorate ingestion for 14 days—a 42% reduction that was statistically significant ($p < 0.01$). The uptake was 27.1% 14 days after cessation of perchlorate. Measurements of uptake at 4 and 8 hr revealed similar degrees of inhibition during ingestion of perchlorate and a similar return to baseline after it was stopped. In the third study, the same investigators then administered 3 mg of potassium perchlorate daily (perchlorate at about 0.03 mg/kg per day for a 70-kg person) to eight healthy men (ages not given) for 14 days (Lawrence et al. 2001). The mean 24-hr thyroid uptake of radioactive iodide was 16.1% at baseline and 14.5% during perchlorate ingestion, and the mean 8-hr uptake values were 13.8% and 11.8%, respectively; neither change was statistically significant. There were no changes in serum T_4, T_3, or TSH concentrations.

In the fourth and most comprehensive study (Greer et al. 2002), 21 healthy women and 16 healthy men (mean age, 38 years; range, 18-57 years) were given potassium perchlorate in doses of perchlorate at 0.007-0.5 mg/kg of body weight per day for 14 days (the daily dose was given in 400 mL of water with instructions that 100 mL be consumed four times each day). The doses were chosen on the basis of the effects observed by Lawrence et al. (2000). Thyroid uptake of radioiodide was measured 8 and 24 hr after radioiodide administration at baseline, on days 2 and 14 days of perchlorate administration, and 15 days later. On day 14, the 24-hr radioiodide uptake was 98.2% of the baseline value in the subjects given 0.007 mg/kg (Table 2-1), a 1.8%, and statistically insignificant, decrease, well within the variation of repeated measurements in healthy subjects. The day-14 24-hr radioiodide uptake value was 83.6% of the baseline value (16.4% decrease) in the subjects given 0.02 mg/kg, 55.3% of the baseline value (44.7% decrease) in those given 0.1 mg/kg, and 32.9% of the baseline value (67.1% decrease) in those given 0.5 mg/kg.

The results of thyroid radioiodide uptake measurements on day 2 of perchlorate administration were very similar to those on day 14 in the three higher dose groups (uptake was not measured on day 2 in the lowest dose group) (Table 2-1), indicating that the effect did not change with time. The 8-hr thyroid radioiodide uptake values, both as percentage of dose and as percentage of baseline, on days 2 and 14 were very similar to those at 24 hr (not shown). The thyroid uptake values 15 days after exposure were very similar to the baseline values, indicating rapid disappearance of inhibition when that had occurred. The results were similar in women and men. The investigators concluded that the no-observed-effect-level (NOEL) for perchlorate-induced inhibition of thyroid iodide uptake was 0.007 mg/kg per day. On the basis of the inhibition of uptake that occurred at higher doses of perchlorate and the potency of perchlorate as an inhibitor of iodide

TABLE 2-1 24-Hour Thyroid Radioiodide Uptake in Healthy Subjects before, during, and after Oral Administration of Potassium Perchlorate for 14 Days

	No. of Subjects	24-Hr Thyroid Uptake (% of Dose) (Mean ± SE)	Uptake as % of Baseline (Mean ± SE)	P, Compared with Baseline
0.007 mg/kg-day				
Baseline	7	18.1 ± 3.1		
Day 14	7	16.5 ± 1.6	98.2 ± 8.3	
Day 15 post-exposure	7	17.3 ± 2.5	100.3 ± 8.4	
0.02 mg/kg-day				
Baseline	10	18.4 ± 1.2		
Day 2	8	15.7 ± 1.4	82.8 ± 5.6	<0.05
Day 14	10	15.2 ± 1.1	83.6 ± 4.1	<0.005
Day 15 post-exposure	10	19.1 ± 1.3	105.3 ± 5.5	
0.1 mg/kg-day				
Baseline	10	19.9 ± 2.1		
Day 2	8	11.8 ± 1.7	59.2 ± 3.5	<0.005
Day 14	10	11.0 ± 1.6	55.3 ± 3.9	<0.005
Day 15 post-exposure	10	20.8 ± 2.2	106.6 ± 9.1	
0.5 mg/kg-day				
Baseline	10	21.6 ± 2.0		
Day 2	8	6.5 ± 0.6	30.6 ± 2.6	<0.005
Day 14	10	6.9 ± 0.9	32.9 ± 3.8	<0.005
Day 15 post-exposure	10	21.7 ± 2.0	104.6 ± 9.4	

Source: Greer et al. 2002.

uptake in vitro, they estimated that a daily intake of 150 μg of iodide would protect against the effects of a daily intake of 4 mg of perchlorate (0.06 mg/kg for a 70-kg person).

Serum perchlorate concentrations ranged from 0.10 to 0.17 μg/mL during administration of 0.1 mg/kg per day and from 0.45 to 0.85 μg/mL during administration of 0.5 mg/kg per day. The serum perchlorate half-life after cessation of this dose averaged 8.1 hr (range, 6.0-9.3); the half-life in the other groups could not be measured, because the values were or became undetectable very soon after perchlorate administration was stopped.

There were no changes in serum T_4, T_3, and TSH concentrations, which were measured repeatedly during the study in 24 subjects, except for a very small decrease in serum TSH concentrations in the subjects given 0.5 mg/kg per day (not an increase, as would be expected if thyroid secretion decreased, and probably explicable on the basis of multiple measurements of TSH concentration, which varies over the course of a day). One woman had a slightly high serum TSH concentration at baseline (18 mU/L), and it was slightly lower (15 mU/L) on day 14 of perchlorate (0.007 mg/kg per day) administration.

In the fifth study, the chronic effects of two doses of potassium perchlorate were evaluated in a small study of 13 healthy subjects (Braverman et al. 2004). Four subjects were given placebo, five were given 0.5 mg of perchlorate daily, and four were given 3 mg of perchlorate daily for 6 months (these doses correspond to 0.007 and 0.04 mg/kg, respectively, for a 70-kg subject). Serum TSH, free T_4 index, and T_3 were measured at baseline, monthly during perchlorate administration, and 1 month after exposure. Thyroid radioiodide uptake was measured at baseline, at 3 and 6 months during perchlorate administration, and 1 month after exposure. There were no changes in any of the serum measurements or in thyroid radioiodide uptake, compared with baseline, at any time in either perchlorate group or in the placebo group.

The results of the studies, in which thyroid function was assessed in several ways, are remarkably consistent. The study subjects were healthy men and women 18-57 years old; none was taking medications that might influence thyroid radioiodide independently of perchlorate. They were free-living, eating a self-selected diet. In the studies in which thyroid radioiodide uptake was measured, the baseline values varied somewhat among the subjects, but no more than expected in healthy people eating their usual diet. The normal range for 24-hr thyroid uptake of radioiodide in many places in the United States is 10-30%, also reflecting variation in dietary iodide intake. Although individual study groups were small, from four to 10 subjects in the studies of thyroid radioiodide uptake, the results were highly consistent within each treatment group in that the variance of the change, or lack of change, in thyroid radioiodide uptake during potassium perchlorate administration was similar to or less than the variance at baseline (Table 2-1). The effects of similar doses of potassium perchlorate on thyroid radioiodide uptake were very similar; a daily perchlorate dose of 0.007 mg/kg had no effect in two studies (Greer et al. 2002; Braverman et al. 2004), a daily dose of 0.02 mg/kg had a small effect (about 15% inhibition of thyroid iodide uptake) in Greer et al. (2002), and daily doses of 0.03 and 0.04 mg/kg had no effect in two other studies (Lawrence et al. 2000;

Braverman et al. 2004). Those results have been analyzed in multiple ways, but the experimental results are clear: in healthy subjects, a dose of perchlorate of 0.007 mg/kg per day has no effect on thyroid radioiodide uptake or any other measure of thyroid function, and doses of 0.02-0.04 mg/kg per day have minimal or no effect on those measures.

Summary of Potential Perchlorate-Induced Perturbations of Thyroid Function in Healthy Humans

The duration of the studies of potassium perchlorate administration in healthy subjects varied from 2 weeks to 6 months. In all those studies, there were no changes in serum T_4, T_3, and TSH concentrations to suggest that there had been any decrease in thyroid hormone secretion. Some doses of perchlorate given for 2 weeks did inhibit thyroid uptake of radioiodide. A high dose given for 4 weeks lowered thyroid iodide content by 25%, indicating some decrease in iodide uptake (Brabant et al. 1992), and resulted in a very small fall in serum free T_4 concentrations, but serum TSH concentrations were also lower—not higher, as would occur if serum free T_4 concentrations decreased to an important extent.

If inhibition of iodide uptake were prolonged, the result should be iodide deficiency in the thyroid, similar to that which occurs in dietary iodide deficiency and with the same consequences and compensation. Lack of iodide in the thyroid gland should cause a decrease in T_4 and T_3 synthesis and secretion and therefore a fall in serum T_4 and T_3 concentrations. TSH secretion would then increase, and, assuming an adequate iodide intake, iodide uptake and T_4 and T_3 synthesis and secretion would return toward normal. Thus, there should be complete compensation if the maximal effect of the dose of perchlorate was only partial inhibition of iodide uptake by the thyroid gland. Even if the block were more complete, substantial or complete compensation would be expected although the thyroid gland might enlarge.

There is therefore little likelihood that the decreases in thyroid iodide uptake found in the short-term studies of perchlorate administration, even at the higher doses, would be sustained, and it is highly likely that iodide uptake would return to normal. Possible exceptions might occur if the dose of perchlorate were very high comparable to those given to patients with hyperthyroidism, or if the person had severe iodide deficiency. Given the compensation that is known to occur in people with iodide deficiency, as discussed earlier, it is highly likely that in people with a normal iodide intake the dose of perchlorate would have to reduce thyroid iodide uptake by at least 75% for a sustained period (several months or longer) for iodide uptake and thyroid hormone production to decline enough to cause adverse health effects (equivalent to reducing dietary iodide intake by 75%). In adults, that is likely to require sustained exposure to more than 30 mg of perchlorate per day (0.4 mg/kg per day for a 70-kg person), on the basis of the clinical studies in healthy subjects and the studies of long-term treatment of hyperthyroidism, both described in this chapter, and the studies of environmental exposure, described in Chapter 3 (Gibbs et al. 1998; Lamm et al. 1999; Crump et al. 2000). In pregnant women, infants and children, and people who have a low iodide intake or pre-existing thyroid dysfunction, the dose required to cause a decrease in thyroid hormone production may be lower. However, a dose that does not inhibit thyroid iodide uptake will not affect thyroid function, even in subjects with an abnormal thyroid gland or a very low iodide intake.

NONTHYROID EFFECTS OF PERCHLORATE

The NIS is present in the salivary glands; mammary glands, especially during lactation; stomach; choroid plexus of the brain; and ciliary body of the eye (Dohan et al. 2003). However, the iodide that is transported into those tissues is not further metabolized, as it is in the thyroid gland, but is rapidly returned to the circulation or secreted into the saliva or breast milk. Furthermore, iodide transport into these tissues is not known to be required for their normal function, with the possible exception of mammary tissue. TSH increases the content of the NIS only in thyroid tissue. Perchlorate acutely inhibits iodide transport in salivary and mammary tissue, but it does not appear to reduce the iodide content of breast milk (see Chapter 3). As noted above, very small amounts of the NIS have been detected in other tissues, including the heart, kidneys, lungs, and placenta.

Perchlorate is not known to cause congenital malformations, but the relationship has not been well studied. The adverse effects of perchlorate given to hyperthyroid patients are described in the preceding section. Note that the effects occurred only in patients with hyperthyroidism given very high doses (many years ago) and that the effects have not been described in any of the more recent studies in which perchlorate was given to patients with hyperthyroidism in lower doses but for periods as long as 2 years.

The possibility that perchlorate will adversely affect the immune system was raised by suggestions that some of the side effects of high doses of perchlorate—rashes, aplastic anemia, or agranulocytosis—might have been immunologic responses. Whether those effects were caused by a direct toxic effect of perchlorate itself or a contaminant of it or by an immunologic reaction to the drug or a contaminant is not known. The fact that the effects were dose-dependent argues for direct toxicity rather than an immunologic reaction.

An immunologic effect of perchlorate might also be suggested by the finding that serum concentrations of the antibodies that stimulate thyroid secretion in patients who have hyperthyroidism caused by Graves disease decline during treatment with antithyroid drugs, including perchlorate (Wenzel and Lente 1984). However, the decline in serum concentrations of those antibodies follows, rather than precedes, the drug-induced decrease in thyroid hormone secretion, and declines also occur in patients treated with other drugs, radioiodide, or thyroidectomy. Regarding a possible immunologic effect of perchlorate, a letter to the editor in 1984 reported that incubation of perchlorate at 0.12 or 1.2 mg/mL with lymphocytes of healthy people for 10 days inhibited immunoglobulin production and lymphocyte transformation (Weetman et al. 1984). However, it is not possible to assess potential clinical effects from experiments in which high doses of perchlorate were added directly to immune cells in vitro. In summary, there is no evidence that regular ingestion of perchlorate in any dose causes immunologic abnormalities in humans.

REFERENCES

Abalovich, M., S. Gutierrez, G. Alcaraz, G. Maccallini, A. Garcia, and O. Levalle. 2002. Overt and subclinical hypothyroidism complicating pregnancy. Thyroid 12(1):63-68.

al-Adsani, H., L.J. Hoffer, and J.E. Silva. 1997. Resting energy expenditure is sensitive to small dose changes in patients on chronic thyroid hormone replacement. J. Clin. Endcrinol. Metab. 82(4):1118-1125.

Allan, W.C., J.E. Haddow, G.E. Palomaki, J.R. Williams, M.L. Mitchell, R.J. Hermos, J.D. Faix, and R.Z. Klein. 2000. Maternal thyroid deficiency and pregnant complications: Implications for population screening. J. Med. Screen. 7(3):127-130.

American Heritage. 2003. Children's Science Dictionary. Houghton-Mifflin Company.

Andersen, S., K.M. Pedersen, I.B. Pedersen, and P. Laurberg. 2001. Variations in urinary iodine excretion and thyroid function. A 1-year study in healthy men. Eur. J. Endocrinol. 144(5):461-465.

Bartalena, L., S. Brogioni, L. Grasso, F. Bogazzi, A. Burelli, and E. Martino. 1996. Treatment of amiodarone-induced thyrotoxicosis, a difficult challenge: Results of a prospective study. J. Clin. Endocrinol. Metab. 81(8):2930-2933.

Bianco, A.C., D. Salvatore, B. Gereben, M.J. Berry, and P.R. Larsen. 2002. Biochemistry, cellular and molecular biology, and physiological roles of the iodothyronine selenodeiodinases. Endocr. Rev. 23(1):38-89.

Biondi, B., E.A. Palmieri, G. Lombardi, and S. Fazio. 2002. Effects of subclinical thyroid dysfunction on the heart. Ann. Intern. Med. 137(11):904-914.

Brabant, G., P. Bergmann, C.M. Kirsch, J. Kohrle, R.D. Hesch, and von zur Muhlen. 1992. Early adaptation of thyrotropin and thyroglobulin secretion to experimentally decreased iodide supply in man. Metabolism 41(10):1093-1096.

Braverman, L.E., X. He, S. Pino, B. Magnani, and A. Firek. 2004. The effect of low dose perchlorate on thyroid function in normal volunteers [abstract]. Thyroid 14(9):691.

Calvo, R.M., E. Jauniaux, B. Gulbis, M. Asuncion, C. Gervy, B. Contempre, and G. Morreale de Escobar. 2002. Fetal tissues are exposed to biologically relevant free thyroxine concentrations during early phases of development. J. Clin. Endocrinol. Metab. 87(4):1768-1777.

Cao, X.Y., X.M. Jiang, Z.H. Dou, M.A. Rakeman, M.L. Zhang, K. O'Donnell, T. Ma, K. Amette, N. DeLong, and G.R. DeLong. 1994. Timing of vulnerability of the brain to iodine deficiency in endemic cretinism. N. Engl. J. Med. 331(26):1739-1744.

Carr, D., D.T. McLeod, G. Parry, and H.M. Thornes. 1988. Fine adjustment of thyroxine replacement dosage: Comparison of the thyrotrophin releasing hormone test using a sensitive thyrotrophin assay with measurement of free thyroid hormones and clinical assessment. Clin. Endocrinol. 28(3):325-333.

Connell, J.M. 1981. Long-term use of potassium perchlorate. Postgrad. Med. J. 57(670):516-517.

Crooks, J., and E.J. Wayne. 1960. A comparison of potassium perchlorate, methythiouracil, and carbimazole in the treatment of thyrotoxicosis. Lancet 1:401-404.

Crump, C., P. Michaud, R. Tellez, C. Reyes, G. Gonzalez, E.L. Montgomery, K.S. Crump, G. Lobo, C. Becerra, and J.P. Gibbs. 2000. Does perchlorate in drinking water affect thyroid function in newborns or school-age children? J. Occup. Environ. Med. 42(6):603-612.

Delange, F.M. 2000. Iodine deficiency. Pp. 295-316 in Werner and Ingbar's The Thyroid: A Fundamental and Clinical Text, 8th Ed., L.E. Braverman, and R.D. Utiger, eds. New York: Lippincott Williams and Wilkins.

Dohan, O., A. De La Vieja, V. Paroder, C. Riedel, M. Artani, M. Reed, C.S. Ginter, and N. Carrasco. 2003. The sodium/iodide Symporter (NIS): Characterization, regulation, and medical significance. Endocr. Rev. 24(1):48-77.

Dunn, J.T., and A.D. Dunn. 2000. Thyroglobulin: Chemistry, biosynthesis and proteolysis. Pp. 91-104 in Werner and Ingbar's The Thyroid: A Fundamental and Clinical Text, 8th Ed., L.E. Braverman, and R.D. Utiger, eds. New York: Lippincott Williams and Wilkins.

Eisenstein, Z., M. Weiss, Y. Katz, and H. Bank. 1992. Intellectual capacity of subjects exposed to methimazole or propylthiouracil in utero. Eur. J. Pediatr. 151(8):558-559.

Fisher, D.A., and R.S. Brown. 2000. Thyroid physiology in the perinatal period and during childhood. Pp. 959-972 in Werner and Ingbar's The Thyroid: A Fundamental and Clinical Text, 8th Ed., L.E. Braverman, and R.D. Utiger, eds. New York: Lippincott Williams and Wilkins.

Foley Jr., T.P. 2000. Congenital hypothyroidism. Pp. 977-983 in Werner and Ingbar's The Thyroid: A Fundamental and Clinical Text, 8th Ed., L.E. Braverman, and R.D. Utiger, eds. New York: Lippincott Williams and Wilkins.

Frank, J.E., J.E. Faix, R.J. Hermos, D.M. Mullaney, D.A. Rojan, M.L. Mitchell, and R.Z. Klein. 1996. Thyroid function in very low birth weight infants: Effects on neonatal hypothyroidism screening. J. Pediatr. 128(4):548-554.

Gibbs, J.P., R. Ahmad, K.S. Crump, D.P. Houck, T.S. Leveille, J.E. Findlay, and M. Francis. 1998. Evaluation of a population with occupational exposure to airborne ammonium perchlorate for possible acute or chronic effects in thyroid function. J. Occup. Environ. Med. 40(12):1072-1082.

Glinoer, D., P. de Nayer, F. Delange, M. Lemone, V. Toppet, M. Spehl, J.P. Grun, J. Kinthaert, and B. Lejeune. 1995. A randomized trial for the treatment of mild iodine deficiency during pregnancy: Maternal and neonatal effects. J. Clin. Endocrinol. Metab. 80(1):258-269.

Greer, M.A., G. Goodman, R.C. Pleus, and S.E. Greer. 2002. Health effects assessment for environmental perchlorate contamination: The dose response for inhibition of thyroidal radioiodine uptake in humans. Environ. Health Perspect. 110(9):927-937.

Haddow, J.E., G.E. Palomaki, W.C. Allan, J.R. Williams, G.J. Knight, J. Gagnon, C.E. O'Heir, M.L. Mitchell, R.J. Hermos, S.E. Waisbren, J.D. Faix, and R.Z. Klein. 1999. Maternal thyroid deficiency during pregnancy and subsequent psychological development in the child. N. Engl. J. Med. 341(8):549-555.

Hennemann, G., R. Docter, E.C. Friesema, M. de Jong, E.P. Krenning, and T.J. Visser. 2001. Plasma membrane transport of thyroid hormones and its role in thyroid hormone metabolism and bioavailability. Endocr. Rev. 22(4):451-476.

Hollowell, J.G., N.W. Staehling, W.H. Hannon, D.W. Flanders, E.W. Gunter, G.F. Maberly, L.E. Braverman, S. Pino, D.T. Miller, P.L. Garbe, D.M. DeLozier, and R.J. Jackson. 1998. Iodine nutrition in the United States. Trends and public health implications: Iodine excretion data from National Health and Nutrition Examination Surveys I and III (1971-1974 and 1988-1994). J. Clin. Endocrinol. Metab. 83(10):3401-3408.

IOM (Institute of Medicine). 2000. Dietary Reference Intakes for Vitamin A, Vitamin K, Arsenic, Boron, Chromium, Copper, Iodine, Iron, Manganese, Molybdenum, Nickel, Silicon, Vanadium, and Zinc. Washington, DC: National Academy Press.

Kilby, M.D, N. Gittoes, C. McCabe, J. Verhaeg, and J.A. Franklyn. 2000. Expression of thyroid receptor isoforms in the human fetal central nervous system and the effects if intrauterine growth restriction. Clin. Endocrinol. 53(4):469-477.

Klein, R.Z., J.E. Haddow, J.D. Faix, R.S. Brown, R.J. Hermos, A. Pulkkinen, and M.L. Mitchell. 1991. Prevalence of thyroid deficiency in pregnant women. Clin. Endocrinol. 35(1):41-46.

Kosugi, S., H. Okamoto, A. Tamada, and F. Sanchez-Franco. 2002. A novel peculiar mutation in the sodium/iodide symporter gene in Spanish siblings with iodide transport defect. J. Clin. Endocrinol. Metab. 87(8):3830-3836.

Kung, A.W., T.T. Lao, M.T. Chau, S.C. Tam, and L.C. Low. 2000. Goitrogenesis during pregnant and neonatal hypothyroxinaemia in a borderline iodine sufficient area. Clin. Endocrinol. 53(6):725-731.

Lamm, S.H., L.E. Braverman, F.X. Li, K. Richman, S. Pino, and G. Howearth. 1999. Thyroid health status of ammonium perchlorate workers: A cross-sectional occupational health study. J. Occup. Environ. Med. 41(4):248-260.

Lawrence, J.E., S.H. Lamm, S. Pino, K. Richman, and L.E. Braverman. 2000. The effect of short-term low-dose perchlorate on various aspects of thyroid function. Thyroid 10(8):659-663.

Lawrence, J., S. Lamm, and L.E. Braverman. 2001. Low dose perchlorate (3 mg daily) and thyroid function. Thyroid 11(3):295.

Liu, H., N. Momotani, J.Y. Noh, N. Ishikawa, K. Takebe, and K. Ito. 1994. Maternal hypothyroidism during early pregnant and intellectual development of the progeny. Arch. Intern. Med. 154(7):785-787.

Mariash, C.N., G.W. Anderson, and J.H. Oppenheimer. 2000. Molecular actions of thyroid hormone. Pp. 174-195 in Werner and Ingbar's The Thyroid: A Fundamental and Clinical Text, 8th Ed., L.E. Braverman, and R.D. Utiger, eds. New York: Lippincott Williams and Wilkins.

Martino, E., S. Mariotti, F. Aghini-Lombardi, M. Lenziardi, S. Morabito, L. Baschieri, A. Pinchera, L.E. Braverman, and M. Safran. 1986. Short term administration of potassium perchlorate restores euthyroidism in amiodarone iodine-induced hypothyroidism. J. Clin. Endocrinol. Metab. 63(5):1233-1236.

Matte, R., L.G. Ste-Marie, R. Comtois, P. D'Amour, A. Lacroix, R. Chartrand, R. Poisson, and C.H. Bastomsky. 1981. The pituitary-thyroid axis after hemithyroidectomy in euthyroid man. J. Clin. Endocrinol. Metab. 53(2):377-380.

Media Lab. 2004. Liver. Clipart. The Media Lab, Department of Biochemistry, UW-Madison, Madison, WI [Online]. Available: http://www.biochem.wisc.edu/medialab/clipart.html [accessed Jan. 18, 2004].

Morgans, M.E., and W.R. Trotter. 1960. Potassium perchlorate in thyrotoxicosis. [Letter]. Br. Med. J. (October 8):1086-1087.

NCHS (National Center for Health Statistics). 2002. Iodine Level, United States, 2000. National Center for Health Statistics, Centers for Disease Control and Prevention, U.S. Department of Health and Human Services, Hyattsville, MD. [Online]. Available: http://www.cdc.gov/nchs/products/pubs/pubd/hestats/iodine.htm [accessed July 10, 2004].

Newnham, H.H., D.J. Topliss, B.A. Le Grand, N. Chosich, R.W. Harper, and J.R. Stockigt. 1988. Amiodarone-induced hyperthyroidism: Assessment of the predictive value of biochemical testing and response to combined therapy using propylthiouracil and potassium perchlorate. Aust. N.Z. J. Med. 18(1):37-44.

Nolte, W., R. Muller, H. Siggelkow, D. Emrich, and M. Hufner. 1996. Prophylactic application of thyrostatic drugs during excessive iodine exposure in euthyroid patient with thyroid autonomy: A randomized study. Eur. J. Endocrinol. 134(3):337-341.

Oden, J., and M. Freemark. 2002. Thyroxine supplementation in preterm infants: Critical analysis. Curr. Opin. Pediatr. 14(4):447-452.

Oppenheimer, J.H., and H.L. Schwartz. 1997. Molecular basis of thyroid hormone-dependent brain development. Endocr. Rev. 18(4):462-475.

Parle, J.V., P. Maisonneuve, M.C. Sheppard, P. Boyle, and J.A. Franklyn. 2001. Prediction of all-cause and cardiovascular mortality in elderly people from one low serum thyrotropin result: A 10-year cohort study. Lancet 358(9285):861-865.

Pharoah, P.O. 1993. Iodine-supplementation trials. Am. J. Clin. Nutr. 57(Suppl. 2):276S-279S.

Pop, V.J., J.L. Kuijpens, A.L. van Baar, G. Verkerk, M.M. van Son, J.J. de Vijlder, T. Vulsma, W.M. Wiersinga, H.A. Drexhage, and H.L. Vader. 1999. Low maternal free thyroxine concentrations during early pregnancy are associated with impaired psychomotor development in infancy. Clin. Endocrinol. 50(2):149-155.

Pop, V.J., E.P. Brouwers, H.L. Vader, T. Vulsma, A.L. van Baar, and J.J. de Vijlder. 2003. Maternal hypothyroxinemia during early pregnancy and subsequent child development: A 3-year follow-up study. Clin. Endcrinol. 59(3):282-288.

Reichert, L.J., and H.A. de Rooy. 1989. Treatment of amiodarone induced hyperthyroidism with potassium perchlorate and methimazole during amiodarone treatment. BMJ 298(6687):1547-1548.

Ron, E., M.M. Doody, D.V. Becker, A.B. Brill, R.F. Curtis, M.B. Goldman, B.S. Harris III, D.A. Hoffman, W.M. McConahey, H.R. Maxon, S. Preston-Martin, M.E. Warshauer, F.L. Wong, and J.D. Boice Jr. 1998. Cancer mortality following treatment for adult hyperthyroidism. Cooperative Thyrotoxicosis Therapy Follow-up Study Group. JAMA 280(4):347-355.

Roti, E., and A.G. Vagenakis. 2000. Effect of excess iodide: Clinical aspects. Pp. 316-329 in Werner and Ingbar's The Thyroid: A Fundamental and Clinical Text, 8th Ed., L.E. Braverman, and R.D. Utiger, eds. New York: Lippincott Williams and Wilkins.

Santiago-Fernandez, P., R. Torres-Barahona, J.A. Muela-Martinez, G. Rojo-Martinez, E. Garcia-Fuentes, M.J. Garriga, A.G. Leon, and F. Soriguer. 2004. Intelligence quotient and iodine intake: A cross-sectional study in children. J. Clin. Endocrinol. Metab. 89(8):3851-3857.

Schlumberger, M.J. 1998. Papillary and follicular thyroid carcinoma. N. Engl. J. Med. 338(5):297-306.

Smit, B.J., J.H. Kok, T. Vulsma, J.M. Briet, K. Boer, and W.M. Wiersinga. 2000.

Neurologic development of the newborn and young child in relation to maternal thyroid function. Acta Pediatr. 89(3):291-295.

Soldin, O.P., L.E. Braverman, and S.H. Lamm. 2001. Perchlorate clinical pharmacology and human health: A review. Ther. Drug Monit. 23(4):316-331.

Soldin, O.P., R.E. Tractenberg, and J.C. Pezzullo. 2005. Thyroxine, thyroid-stimulating hormone, and urinary iodine associations in pregnant and nonpregnant women of childbearing age in the US. Ther. Drug Monit. Clin. Toxicol. 27(2):178-185.

Stanbury, J.B., and J.B. Wyngaarden. 1952. Effect of perchlorate on the human thyroid gland. Metabolism 1(6):533-539.

Staub, J.J., B.U. Althaus, H. Engler, A.S. Ryff, P. Trabucco, K. Marquardt, D. Burckhardt, J. Girard, and B.D. Weintraub. 1992. Spectrum of subclinical and overt hypothyroidism: Effect on thyrotropin, prolactin, and thyroid reserve, and metabolic impact on peripheral target tissues. Am. J. Med. 92(6):631-642.

Surks, M.I., E. Ortiz, G.H. Daniels, C.T. Sawin, N.F. Col, R.H. Cobin, J.A. Franklyn, J.M. Hershman, K.D. Burman, M.A. Denke, C. Gorman, R.S. Cooper, and N.J. Weissman. 2004. Subclinical thyroid disease: Scientific review and guidelines for diagnosis and treatment. JAMA 291(2):288-338.

Taurog, A.M. 2000. Hormone synthesis: Thyroid iodine metabolism. Pp. 61-85 in Werner and Ingbar's The Thyroid: A Fundamental and Clinical Text, 8th Ed., L.E. Braverman, and R.D. Utiger, eds. New York: Lippincott Williams and Wilkins.

Tiwari, B.D., M.M. Godbole, N. Chattopadhyay, A. Mandal, and A. Mithal. 1996. Learning disabilities and poor motivation to achieve due to prolonged iodine deficiency. Am. J. Clin. Nutr. 63(5):782-786.

Vanderpump, M.P., W.M. Tunbridge, J.M. French, D. Appleton, D. Bates, F. Clark, J. Grimley Evans, D.M. Hasan, H. Rodgers, F. Tunbridge, et al. 1995. The incidence of thyroid disorders in the community: A twenty-year follow-up of the Whickham Survey. Clin. Endocrinol. 43(1):55-68.

Vanderpump, M.P., W.M. Tunbridge, J.M. French, D. Appleton, D. Bates, F. Clark, J. Grimley Evans, H. Rodgers, F. Tunbridge, and E.T. Young. 1996. The development of ischemic heart disease in relation to autoimmune thyroid disease in a 20-year follow-up study of an English community. Thyroid 6(3):155-160.

van den Briel, T., C.E. West, N. Bleichrodt, F.J. van de Vijver, E.A. Ategbo, and J.G. Hautvast. 2000. Improved iodine status is associated with improved mental performance of schoolchildren in Benin. Am. J. Clin. Nutr. 72(5):1179-1185.

Van Sande, J., C. Massart, R. Beauwens, A. Schoutens, S. Costagliola, J.E. Dumont, and J. Wolff. 2003. Anion selectivity by the sodium iodide symporter. Endocrinology 144(1):247-252.

van Wassenaer, A.G., J.H. Kok, J.J. de Vijlder, J.M. Briet, B.J. Smit, P. Tamminga, A. van Baar, F.W. Dekker, and T. Vulsma. 1997. Effects of thyroxine supplementation on neurologic development of infants born at less that 30 weeks' gestation. N. Engl. J. Med. 336(1):21-26.

Vulsma, T., M.H. Gons, and J.J. de Vijlder. 1989. Maternal-fetal transfer of thyroxine in congenital hypothyroidism due to a total organification defect or thyroid agenesis. N. Engl. J. Med. 321(1):13-16.

Weetman, A.P., C. Gunn, R. Hall, and A. McGregor. 1984. Immunosuppression by perchlorate. Lancet 1(8382):906.

Wenzel, K.W., and J.R. Lente. 1984. Similar effects of thionamide drugs and perchlorate on thyroid-stimulating immunoglobulins in Graves' disease: Evidence against an immunosuppressive action of thionamide drugs. J. Clin. Endocrinol. Metab. 58(1):62-69.

WHO (World Health Organization). 1996. Iodine. Pp. 49-71 in Trace Elements in Human Nutrition and Health. Geneva: World Health Organization.

Wolff, J. 1998. Perchlorate and the thyroid gland. Pharmacol. Rev. 50(1):89-105

Yale New Haven Health. 2004. Health Library: Thyroid. Thyroid Image. [Online]. Available: http://yalenewhavenhealth.org/library/healthguide/en-us/support/topic.asp?hwid=tp10822 [accessed November 17, 2004].

3

Epidemiologic Studies of Occupational and Environmental Exposures to Perchlorate

THIS chapter considers observational epidemiologic studies of perchlorate exposures and measures of thyroid function and thyroid diseases in selected populations. The committee considered published and unpublished epidemiologic data in forming its conclusions regarding the possible effects on thyroid function of exposure to perchlorate and in assessing the relevant exposure limits. More weight was given to published reports, and unpublished data were considered only when sufficient information was made available to the committee to evaluate the methods used. Most of the unpublished data considered were presented to the committee in open meetings. The committee considered one unpublished master's thesis (Schwartz 2001). In weighing the evidence on perchlorate exposure, the committee emphasized studies with the soundest scientific methods and studies that included biologically sensitive groups, such as pregnant women, fetuses, and neonates. Of the epidemiologic studies, one is an unpublished mortality study of workers at a chemical plant who had simultaneous exposures to many agents, including perchlorate (Rockette and Arena 1983); three are cross-sectional studies of occupational cohorts that had respiratory exposures to perchlorate and for which assessments of thyroid function were obtained (Gibbs et al. 1998; Lamm et al. 1999; Braverman et al. 2004); and the remainder are primarily ecologic investigations that compared measures of thyroid function or disease and neurodevelopmental outcomes in neonates, children, and adults in geographically defined groups with and without detectable perchlorate in community water supplies (Lamm and Doemland 1999; Brechner et al. 2000; Crump et al. 2000; F.X.

Li et al. 2000; Z. Li et al. 2000; Li et al. 2001; Schwartz 2001; Chang et al. 2003; Kelsh et al. 2003; Lamm 2003; Buffler et al. 2004; Gibbs 2004a,b). The ecologic study of Morgan and Cassady (2002) assessed cancer incidence in a community exposed to both perchlorate and trichloroethylene in drinking water. No epidemiologic studies, either published or unpublished, have measured both thyroid outcomes and perchlorate exposure from drinking water in the same people.

Many of the epidemiologic data related to the effects of perchlorate are derived from ecologic studies. The smallest units on which exposure or outcome data are available are geographically defined units, most commonly counties, states, or countries; exposure data, outcome data, or both are available only at that level, not on individual subjects. Because ecologic studies do not include information about exposure and outcome in individuals, they are considered to be the weakest type of observational studies. In ecologic studies, comparisons are made between exposures and outcomes among large units. An example of an ecologic study would be one that compares the mean serum concentration of thyroid-stimulating hormone (TSH) in infants born in a county in which the drinking-water supply contains perchlorate with the mean serum TSH concentration in infants born in a county in which the drinking water does not contain perchlorate.

That design is subject to what is referred to as the ecologic fallacy: associations observed at the ecologic level may not apply at the individual level. For example, an observation that the average serum TSH concentration is higher in newborns in a city in which the water supply contains perchlorate than in newborns in cities in which the drinking water does not contain perchlorate would be compatible with a perchlorate effect on thyroid function. In that example, being compatible with an association assumes that no other important variations between the cities could account for the difference in serum TSH concentrations. However, it is not known from such a study whether infants who have high TSH concentrations were themselves exposed to perchlorate during gestation. How well ecologic studies are able to characterize individual exposure depends, in part, on how much variability of exposure there is in the geographic unit. For example, ecologic studies of the correlation of dietary fat intake and breast-cancer mortality by country would undoubtedly suffer from considerable variation in the average dietary fat intake among the persons in a given country. The use of average community exposures for ecologic studies of perchlorate in municipal drinking water may be less subject to error, depending on the population coverage of the water supply. In many of the exposed communities included in the literature, an entire community's water supply has measurable perchlorate. Thus, it is likely that everyone has some exposure to perchlorate, depending on how much of their water intake comes from the community's water supply. Individual variation in water exposure undoubtedly still occurs, however, because of the use of wellwater or bottled water or because of nonuniform distributions of contaminants in a geographic area. Some degree of individual variability undoubtedly exists, but it may not be as great as for other types of exposures that are not part of a communitywide, common-source exposure.

Another limitation of ecologic studies is that their design cannot control for many confounding factors, because such data are not usually available at the population level. Results of ecologic studies can be useful in providing supporting data on a possible causal relation, but they cannot themselves provide direct evidence of causation.

The available pertinent occupational and epidemiologic studies are summarized in Table 3-1 and discussed in the following sections.

STUDIES IN OCCUPATIONAL COHORTS AND ADULTS

An early study of an occupational cohort by Rockette and Arena (1983) reported mortality patterns for 59 selected causes of death among workers at the Niagara Plant of Hooker Chemical. It included people with at least 1 year of employment from January 1, 1949, to December 31, 1978. The cohort consisted of 3,963 workers (3,715 men and 248 women) at the plant in Niagara Falls, New York. Only 13 deaths were recorded among women, and detailed analyses of the female data were not conducted because the sample was so small. The following results are related to men.

Results showed statistically significant excess mortality from stomach cancer (standardized mortality ratio [SMR], 178.9; $p < 0.05$) and respiratory cancers (SMR, 145.9; $p < 0.01$). In the respiratory-cancer category, the SMR was significantly increased for cancer of the lung (SMR, 142.4; $p < 0.01$). A review of work areas and production indicated that exposure to magnesium perchlorate occurred in 1970-1976 with simultaneous exposure to dozens of other chemicals at the plant. Of the work areas in the plant, 12 departments were identified as having exposure to "groups of chemicals." The department referred to as "area 4" included exposure to magnesium perchlorate and 23 other chemicals. Job titles in the department were not provided, so specific jobs were not assessed with regard to exposure to magnesium perchlorate. A comparison of department-specific

TABLE 3-1 Summary of Epidemiologic Studies[a]

Reference	Study Design	Population	Exposure	Adjustment Factors	Outcomes	Major Findings	Comments
			Studies of Occupational Groups and Adult Cohorts				
Rockette and Arena 1983	SMR study, 1970-1978	3,963 chemical workers, 94% male	Magnesium perchlorate, unknown amount, in combination with 23 other chemicals	Age, calendar time	Cause-specific mortality	For males 1970-1978, all cancer SMR = 143.7, $p < 0.01$; respiratory cancer SMR = 169.6, $p < 0.01$; stomach cancer SMR = 236.7, $p < 0.05$	Very few deaths recorded in females, so only data on males analyzed in detail; multiple chemical exposures; no adjustment for cigarette-smoking or potential confounders other than age
Gibbs et al. 1998	Repeated cross-sectional (preshift and postshift)	18 workers exposed to ammonium perchlorate and 83 nonexposed workers	Respiratory exposure at 0.2-436 µg/kg, depending on work area	Duration of shift	Compared preshift and postshift serum free T_4 index, total T_4, and TSH	No significant relation of perchlorate dose to change in thyroid hormone and TSH measures across shift	Respiratory exposure; small number of exposed workers; subject to potential bias of affected workers who left employment before study
Gibbs et al. 1998	Historical cohort	53 high dose, 44 low dose, 192 nonexposed	Lifetime cumulative exposure to perchlorate; mean in "low-dose" group, 3,500 µg/kg; mean "high-dose," 38,000 µg/kg	None	Compared medical records on thyroid hormone and TSH measures and disease; medical surveillance data from cohort	Neither estimated cumulative lifetime exposure nor exposure category significantly associated with abnormalities of thyroid hormone measures and TSH	Respiratory exposure; subject to potential bias of affected workers who left employment before study; no statistical adjustment for BMI or activity level

Lamm et al. 1999	Cross-sectional	31 ammonium perchlorate workers and 21 azide workers at the same site; ages 20-56 yr	Absorbed perchlorate categorized as none, low, medium, and high (mean, 1, 4, 11, 34 mg/shift)	None	Serum TSH, T_4, and T_3; free T_4 index; thyroid hormone binding ratio; thyroid peroxidase antibodies; urinary iodide and creatinine; clinical examination	No significant differences in any outcome measure between perchlorate-exposed and azide-exposed workers; no clinical evidence of thyroid abnormalities in perchlorate-exposed workers	Respiratory exposure; subject to potential bias of affected workers who left employment before study; small number in study groups reduces ability to detect differences
Li et al. 2001	Ecologic	Medicaid enrollees in Nevada by county of residence, 1997-1998	Residence in Clark County (drinking-water mean = 8.9 ppb), Washoe County, or remaining counties combined (nonexposed)	None	Medicaid payment diagnosis of one of nine thyroid diseases or thyroid cancer	Prevalence of thyroid diseases or thyroid cancer not significantly higher in Clark County than in either Washoe or all other nonexposed counties combined	Most thyroid diseases were uncommon, so numbers were small in county comparisons; no adjustment for differences among three county groups in age, sex, race or other potential confounders
Morgan and Cassady 2002	Ecologic	Residents of 13 contiguous census tracts in Redlands, CA (San Bernardino County), 1988-1998	Drinking-water exposure to TCE (0.09-97 ppb) measured since 1980 and ammonium perchlorate (5-98 ppb) since 1997	Age, sex, race, or ethnicity and calendar time	O/E numbers of cancer cases; total and site-specific and separately for children < 15 yr old	SIR for thyroid cancer = 1.0 (95% CI = 0.63-1.47); cancers of colon or rectum and lung or bronchus had significantly lower SIRs and melanoma and uterine cancer had significantly higher SIRs; no cancers were observed more often than expected in children	Timing and duration of exposure to perchlorate unclear; exposure to TCE also present; no adjustment for other potentially confounding variables; expected numbers were based on surrounding four counties and included study county data

TABLE 3-1 *(Continued)*

Reference	Study Design	Population	Exposure	Adjustment Factors	Outcomes	Major Findings	Comments
Braverman et al. 2004	Repeated cross-sectional (pre- and postshift)	29 workers at ammonium perchlorate production facility and 12 nonworker volunteers	Respiratory exposure to ammonium perchlorate; mean absorbed dose per shift 0.3 mg/kg	None	Serum TSH, T_4, T_3, free T_4 index, 14-hr radioactive iodide uptake (RAIU), urinary iodide excretion, serum perchlorate, thiocyanate and nitrate; thyroid size	Decrease in mean RAIU postshift compared to preshift and increase in iodide excretion postshift; serum perchlorate undetectable after 3 days off work but elevated postshift; TSH unchanged, T_4, and T_3, free T_4 index showed small but significant increases postshift	Pre- and postshift evaluation allows for comparison within the same individuals; data are preliminary; subject to potential bias of affected workers leaving employment prior to study; small number of nonexposed volunteers
Studies of Neonatal and Pediatric Populations							
Lamm and Doemland 1999	Ecologic SMR	Clark County NV; six southern CA counties; 700,000 newborns screened 1996-1997	Drinking water; NV, 4-16 ppb continuous; CA, 5-8 ppb sporadic	Hispanic ethnicity	Congenital hypothyroidism	O/E cases = 1.0 (0.90-1.16)	Exposure only sporadic in CA; questionable comparison rates used; not adjusted for birthweight
Brechner et al. 2000	Ecologic	Newborns in Yuma, AZ (1,099) and newborns in Flagstaff, AZ (443) 1994-97	~ 6 ppb in Yuma drinking water; none in Flagstaff water	Race or ethnicity and age at sample collection	Mean log TSH and T_4 in serum	Adjusted mean log TSH significantly higher in Yuma than in Flagstaff; no significant difference in T_4	Not adjusted for birthweight or gestational age; 1999 perchlorate used to characterize the study period, 1994-1997

Crump et al. 2000	Ecologic	6- to 8-yr olds in three cities in Chile, 1999; newborns in the three cities, 1996-1999	Drinking water; Antofagasta, no detectable; Chanaral, 5-7 ppb; Taltal, 100-120 ppb	Age, sex, urinary iodide	Serum TSH, T_4, free T_4, T_3, liver and kidney function, urinary iodide and creatinine; clinical examination for goiter; family history of goiter by questionnaire	For life-long residents, no significant differences in TSH; adjusted free T_4 significantly higher in Chanaral and Taltal than in Antofagasta; no significant differences in prevalence of goiter in children (~20% in each city); family history goiter ~5 times as high in Taltal; log TSH significantly lower in newborns from Taltal	Drinking-water perchlorate measured in sample from faucets used by children studied; newborn TSH not adjusted for birthweight or gestational age
Z. Li et al. 2000	Ecologic	Newborns in Las Vegas, NV (17,308), newborns in Reno, NV (5,882), 1998-1999	Las Vegas drinking water (9-15 ppb for 7 mo and undetectable for 8 mo; estimated cumulative exposure in pregnancy, 0.9-4.2 mg); Reno (nonexposed)	Sex, age at sample collection, birthweight	Mean T_4 in newborn blood	No significant difference in adjusted mean T_4 concentrations between newborns in Las Vegas and newborns in Reno; no difference in prevalence of low T_4 ($\leq$10th percentile of birth-cohort distribution)	Excluded birthweights <2,500 g and >4,500 g
F.X. Li et al. 2000	Ecologic	Newborns in Las Vegas, NV (407) and Reno, NV (133), 1998-1999	Las Vegas drinking water (9-15 ppb for 7 mo and undetectable for 8 mo; estimated cumulative exposure in pregnancy, 0.9-4.2 mg); Reno (nonexposed)	Age at sample collection, sex	Mean log TSH in newborn blood	No significant difference in adjusted mean log TSH concentrations between newborns in two cities	Excluded birthweights <2,500 g and >4,500 g; adjustment for age at specimen collection dichotomous

TABLE 3-1 *(Continued)*

Reference	Study Design	Population	Exposure	Adjustment Factors	Outcomes	Major Findings	Comments
Schwartz 2001	Ecologic	507,982 CA newborns, 1996	Drinking water; estimates based on averages obtained from 1997 testing of samples from set of state's water systems; categorized as none, low (1-2 ppb), medium (3-12 ppb), high (>12 ppb)	Sex, ethnicity, multiple birth, birthweight, age at specimen collection	Newborn blood screening T_4 and TSH, presumptive and confirmed congenital hypothyroidism	Adjusted mean T_4 significantly lower and mean log TSH significantly higher with increasing perchlorate exposure; odds of presumptive congenital hypothyroidism significantly increased in low and medium exposure categories, but not in high; confirmed congenital hypothyroidism not related to perchlorate exposure	Potential misclassification of perchlorate exposure due to incomplete measurements; failure to account for multiple laboratories doing newborn screening tests
Chang et al. 2003	Ecologic	NV Medicaid recipients, 1996-2000, <18 yr old with diagnosis of ADHD or autism	County of residence; Clark County (Las Vegas) (exposed; median 23.8 ppb), Washoe County (Reno) (nonexposed), all remaining counties (nonexposed)	None	Proportion of Medicaid recipients with diagnosis of ADHD or autism and national percentile rankings for fourth-grade proficiency test scores	No significant differences between exposed and nonexposed counties in the proportion of Medicaid recipients <18 yr old with diagnosis of ADHD or autism; no significant differences in fourth-grade performance scores	Analysis does not account for changes in residence; validity of diagnoses not established; exposure time may not have been biologically relevant; diagnoses made by multiple health-care providers without use of standard criteria; no statistical adjustment for potential confounding variables

Kelsh et al. 2003	Ecologic	Newborns in two CA communities, 1983-1997; Redlands and Mentone (15,330), San Bernardino and Riverside (696,530)	Drinking water; Redlands and Mentone (exposed, up to 9 ppb; mean 1 ppb), San Bernardino and Riverside (nonexposed)	Age at specimen collection, sex, race, or ethnicity, birthweight, multiple births, cal-endar year	Primary congenital hypothyroidism and high serum TSH (mostly >25 μU/mL, sometimes >16 μU/mL)	Residence in exposed community not associated with higher prevalence of congenital hypothyroidism (SPR = 0.45, 95% CI = 0.06-1.64) or high serum TSH in newborns (SPR = 0.72, 95% CI = 0.28-1.54)	Analyses based on only two cases of congenital hypothyroidism and six of high TSH in exposed community; exposure data from single year used to characterize entire interval
Lamm 2003	Ecologic	Newborns in Yuma, San Luis, and Somerton (Yuma County), and Flagstaff, AZ September 1994-June 1998	Drinking water; Yuma exposed; San Luis, Somerton, Flagstaff nonexposed	None	Median serum TSH	Median serum TSH concentration higher in Yuma (20.8 mU/L) and San Luis and Somerton (21 mU/L) than in Flagstaff (14.5 mU/L); newborns in Yuma have exposure to perchlorate, those in San Luis and Somerton do not	Not adjusted for differences in birthweight, gestational age, ethnicity
Buffler et al. 2004	Ecologic	Newborns in CA, 1998 (342,257)	Drinking water, 1997-1998; exposed communities had mean perchlorate >5 ppb; nonexposed communities had perchlorate <5 ppb	Age at specimen collection, sex, ethnicity, birthweight	Primary congenital hypothyroidism and high serum TSH (generally >25 μU/mL)	No statistically significant relation between residence in exposed community and prevalence or odds of primary congenital hypothyroidism or high TSH	Study was extension of Kelsh et al. (2003) addressing lack of exposure assessment for complete study period and small number of cases, although still lacking exposure assessment of individuals; in Buffler et al. (2004), exposure assessment corresponded to newborn screening data period; 15 cases of primary congenital

TABLE 3-1 *(Continued)*

Reference	Study Design	Population	Exposure	Adjustment Factors	Outcomes	Major Findings	Comments
							hypothyroidism and 147 cases of high serum TSH observed

[a]Includes published papers and final study reports.

Abbreviations: ADHD, attention-deficit-hyperactivity disorder; AZ, Arizona; BMI, body-mass index; CA, California; CI, confidence interval; g, gram; mg, milligram; μg/kg, microgram per kilogram; μU/mL, microunits per milliliter; mU/L, milliunits per liter; mo, month; NV, Nevada; O/E, observed over expected; ppb, parts per billion; SIR, standardized incidence ratio; SMR, standardized mortality ratio; SPR, standardized prevalence ratio; T_3, triiodothyronine; T_4, thyroxine; TCE, trichloroethylene; TSH, thyroid-stimulating hormone; yr, year.

mortality of workers in the plant with U.S. and local Niagara County mortality by calendar year found no statistically significant excess mortality for 1970-1978; magnesium perchlorate exposure occurred in "area 4" of the Niagara Falls plant in 1970-1976. However, analyses of the entire cohort by calendar period showed a statistically significant excess of deaths from malignant neoplasms as a group (SMR, 143.7; $p < 0.01$) and for respiratory cancers (SMR, 169.6; $p < 0.01$) and stomach cancers (SMR, 236.7; $p < 0.05$) for 1970-1978. Because of the multiple exposures of those workers and the inability to adjust for important confounding variables, such as cigarette smoking, it was not possible to determine whether perchlorate exposure itself was related to any increase in SMRs.

In another study of occupational exposure among 254 employees at an ammonium perchlorate production facility in Nevada, those who agreed to participate in the study were investigated for possible acute or chronic effects of perchlorate exposure on thyroid function (Gibbs et al. 1998). Two analyses were conducted: a single-shift study and a working-lifetime study. For the single-shift study, differences in the triiodothyronine (T_3) resin uptake assay, the free thyroxine (T_4) index, total T_4, and TSH in serum obtained before and after a work shift were compared in exposed workers ($n = 18$) and control workers ($n = 83$), defined as persons who had not worked in the production areas for the preceding 30 days. Exposures to ammonium perchlorate in this production facility were through the respiratory tract and were estimated by personal breathing-zone sampling and air sampling of work areas in the plant. Full-shift, personal breathing-zone filters in closed-face cassettes were used to characterize the average airborne exposure to ammonium perchlorate in each work area, which was estimated at 0.2-436 μg/kg (mean, 36 μg/kg) per day, depending on work area. Exposure estimates were obtained for the exact shift in which a worker had volunteered for thyroid hormone and TSH measurements to be taken and were adjusted downward by 65% for areas in which workers wore respiratory protection. Workers taking thyroid medications (all of whom were potential controls) were excluded from analyses.

Cumulative working-lifetime exposure estimates were also calculated (Gibbs et al. 1998). Estimates of lifetime exposure of individual workers were based on the product of mean exposures for their work areas, the number of years they worked in those areas, and 2,000 (the estimated number of hours worked per year). Work histories used in the calculation of cumulative lifetime exposures were obtained by reviewing personnel records and interviewing workers. In addition to the individual exposure estimates, three exposure categories were arbitrarily created for the cumulative-exposure analysis; controls, "low" exposure (cumulative lifetime exposure, 500-7,000 μg/kg; mean, 3,500 μg/kg), and "high" exposure (cumulative lifetime exposure, 8,000-88,000 μg/kg; mean, 38,000 μg/kg). Duration of employment of the "low" and "high" exposure groups combined ranged from 1 to 27 years (mean, 8.3 years). Absolute cumulative-exposure estimates by individual and by exposure group were examined separately in relation to thyroid hormone and TSH test results. The thyroid hormone and TSH measures assessed in the cumulative-exposure analysis were the same as those examined in the single-shift analysis. Rather than being contemporaneously obtained, however, thyroid hormone and TSH measures were obtained from reports of routine medical surveillance in the workers' medical records. Those routine medical-surveillance profiles included blood tests of liver, kidney, and bone marrow function, which were also examined in the analysis of cumulative exposure data. Multiple regression analyses were used to compare thyroid hormone and TSH measures with exposure dose estimates, controlling for race, sex, and age for both single-shift and cumulative assessments and, for the single-shift assessments, controlling also for hours awake before the preshift test, number of hours slept in the period before testing, time of day of test, and shift length. Results were not controlled for BMI or activity level.

Results from the single-shift analysis indicated no significant associations between estimated exposure and alterations in any of the thyroid hormone and TSH measures (Gibbs et al. 1998). Only duration of shift was related to serum TSH: the postshift mean serum TSH concentration was higher than the preshift concentration and exceeded it by more after a 12-hr shift than after an 8-hr shift, a difference expected because there is a circadian increase in serum TSH concentrations. In the analysis of cumulative lifetime exposure to ammonium perchlorate, neither the dose estimate nor the categorical exposure variable was significantly related to measures of thyroid hormone and TSH production. They were unrelated to all but one (white blood cell count) of the measures of bone marrow, liver, or kidney function. The average number of years of exposure to ammonium perchlorate in the working-lifetime study was 8.3 (range, 1-27 years); the results of these analyses appear to reflect the effects of chronic exposure in these workers. The working-lifetime data are based, however, only on employees still working at the plant when the study was done; they do not account for those who may have left employment because of thyroid disease or other disease or toxicity. One of the primary limitations of both the single-shift and cumulative exposure analyses is that the route of exposure to perchlorate was inhalation and thus may not reflect health effects of ingestion of drinking water. The authors argue, however, that because ammonium perchlorate is very water-soluble, it is likely that a high percentage of the inhaled dose is absorbed through the respiratory tract. Urinary measures of perchlorate obtained from workers in a later study (see below) indicated significant absorption of perchlorate by the respiratory tract. Other limitations include the small sample size, particularly of the exposed group, which minimizes the statistical power to detect a meaningful difference; and the low participation rate, which raises the possibility that selection bias will be increased by the limitation of studying only current workers.

Lamm et al. (1999) compared the thyroid health status of 37 workers at an ammonium perchlorate production plant in Iron County, Utah, with that of 21 workers involved in the production of sodium azide at the same industrial site. The workers were 20-56 years old. Exposure of workers to perchlorate at both sites was estimated by measuring air concentrations of total and respirable perchlorate particles (exposures to airborne perchlorate ranged from 0.004 to 167 mg/day), and job assignments were categorized as "low," "medium," and "high" exposure on the basis of visible dust generated in the production area. The mean absorbed perchlorate was about 1, 4, 11, and 34 mg/shift for the azide-exposed and low, medium, and high perchlorate-exposure groups, respectively. The validity of that categorization for capturing variable exposures was demonstrated by comparisons of creatinine-adjusted median concentrations of absorbed perchlorate measured in urine samples collected before and after the work shift. Questionnaires were administered that included information on alcohol and tobacco use, medications, and family history of diseases (diabetes, hypertension, rheumatoid arthritis, thyroid disease, and cancer). Thyroid status was assessed by measuring serum TSH, T_4, T_3, free T_4 index, thyroid hormone binding ratio, and thyroid peroxidase antibodies and by conducting clinical examinations. Postshift blood specimens were obtained for a complete blood count and chemistry panel, and urinary iodide and creatinine concentrations were measured to determine whether workers had adequate dietary iodide intake. No statistically significant differences were observed in any of the measures of thyroid status between the azide workers and any of the three perchlorate-exposure groups. No clinical evidence of thyroid abnormalities or evidence of hematotoxicity was observed in any perchlorate-exposure group. The sample sizes in each group were small, however, so it was difficult to detect meaningful differences between exposure groups. This study and that of Gibbs et al. (1998) were cross-sectional investigations of worker populations. Thus, they are not able to account for any effects of exposure that might have occurred in workers who have left employment for any reason. If perchlorate-exposed workers who developed thyroid disease retired or changed jobs because of their illness, any association of exposure with adverse outcomes would be underestimated or missed altogether.

Preliminary analyses of data from another study of workers in the Utah ammonium perchlorate production facility were presented at a public meeting of the committee in May 2004 and later in a draft manuscript (Braverman et al. 2004). Twenty-nine of 40 workers employed for at least 2 years in ammonium perchlorate production were assessed at two times: after 3 days off work (preshift) and during the last of three 12-hr shifts in the plant (postshift). Exposure to perchlorate was respiratory. The authors also examined 12 nonrandomly selected community volunteers who were not working in the plant. Measures included serum T_4, free T_4 index, total T_3, thyroxine-binding globulin, and TSH; 14-hr radioactive iodide uptake (RAIU); urinary iodide excretion; and serum perchlorate, thiocyanate, and nitrate. The perchlorate doses absorbed over a shift averaged 0.3 mg/kg. Mean postshift RAIU (13.5%) was lower than preshift RAIU (21.5%) but preshift RAIU values were higher than those in the nonexposed controls, suggesting upregulation of iodide transport activity. An increase in urinary iodide excretion, 230 μg/g of creatinine (postshift) vs 148 μg/g of creatinine (preshift), was concomitant with postshift RAIU reduction. Serum perchlorate was not detectable after 3 days off work, but was increased in the postshift sample (850 ppb [850 μg/L]). Serum TSH and thyroxine-binding globulin concentrations were within the normal range and were similar in all three comparisons (preshift and postshift workers and nonexposed volunteers). Serum T_4 (8.3 vs 7.7 μg/dL), free T_4 index (2.4 vs 2.2), and total T_3 (147 vs 134 ng/dL) concentrations were significantly higher in postshift than in preshift samples. Thyroid volumes and patterns assessed by ultrasonography were not significantly different between workers and volunteers. The data suggest small acute changes in some measures of thyroid function that are in an opposite direction to an antithyroid effect of perchlorate. No apparent long-term effects on thyroid size or function were observed. As previously mentioned, the analyses are subject to potential bias because of loss from the workforce of workers whose thyroid function has been adversely affected; thus, the study group may have been "selected" for the absence of effects. Because of the preliminary nature of the data, the committee was not able to consider the results of the study in arriving at its conclusions.

Li et al. (2001) analyzed Nevada Medicaid data for the period January 1, 1997, through December 31, 1998, to compare the prevalence of specific thyroid diseases among Medicaid recipients by county of residence as characterized by the presence or absence of perchlorate in public drinking water. Clark County, which includes Las Vegas, is the only Nevada county in which perchlorate was detected in public drinking water during the time of the study; all other counties were considered to be "nonexposed." Data on the size of the Medicaid-eligible population in each county were provided by the state. The prevalence proportions for each of the specific thyroid diseases were calculated for eight county categories (seven individual counties and all other counties combined). Prevalence proportions were then compared among three county groupings: Clark County, with perchlorate in drinking water at 4.1-14 ppb (mean, 8.9 ppb [8.9 μg/L]); Washoe County, which includes Reno, with no detectable perchlorate; and all other counties in the state combined, also with no detectable perchlorate. The thyroid diseases studied were simple and nonspecified goiter, nontoxic nodular goiter, thyrotoxicosis with or without goiter, congenital hypothyroidism, acquired hypothyroidism, thyroiditis, other disorders of the thyroid, and malignant neoplasms of the thyroid (ICD-9 Codes 193 and 240-246). Residence as noted in the Medicaid database was used to determine exposure status (county of residence).

Most of the thyroid diseases studied were uncommon, with a 2-year prevalence in Nevada ranging from 1 in 10,000 for congenital hypothyroidism to 121 in 10,000 for acquired hypothyroidism (Li et al. 2001). There were no statistically significant differences in the prevalence of any thyroid disease between Clark County (the "exposed" county) and Washoe County ("nonexposed" and urban) or between Clark County and all remaining counties combined ("nonexposed" and rural). The prevalence ratio for thyroid cancer in Clark County (28 cases in 122,519, or 2 per 10,000 Medicaid-eligible people) versus Washoe County (nine cases in 29,622, or 3 per 10,000 Medicaid-eligible people) was 0.75 (95% confidence interval [CI], 0.35-1.59), indicating no significant increase in thyroid cancer associated with exposure to perchlorate at the concentrations reported for those areas. No adjustment of prevalence proportions for differences in sex or age distributions among the three geographic areas could be made, because the pertinent data were not available in this ecologic study. In the absence of the ability to adjust for potential confounding variables, comparisons between Clark and Washoe Counties are probably the most informative in that both counties include large metropolitan areas. Clark County had about 4 times more Medicaid enrollees than did Washoe County at the time of the study, so prevalence estimates from Clark County are more stable. The infrequency of some of the thyroid diseases resulted in small numbers of cases, which may have accounted for the failure to achieve statistical significance in several of the comparisons.

In another ecologic study, Morgan and Cassady (2002) compared the observed and expected numbers of incident cancer cases among residents of 13 contiguous census tracts in Redlands, California, in San Bernardino County. That area is served by the Desert Sierra Cancer Surveillance Program, a regional cancer registry that includes four counties in Southern California, one of which is San Bernardino County. All cancers in the region have been reportable to the California cancer registry by law since 1988. Testing of the drinking water from about 20 wells serving Redlands for trichloroethylene (TCE) was initiated in 1980 and for ammonium perchlorate in 1997. Perchlorate in the wells in 2001 was reported at 5-98 ppb (5-98 μg/L), with drinking-water concentrations not exceeding 18 ppb (18 μg/L). TCE in the wells initially ranged from 0.09 to 97 ppb (0.09-97 μg/L) but after water treatment or removal of highly contaminated wells from service has not exceeded 5 ppb (5 μg/L) in drinking water since 1991. Thus, residents of the 13 census tracts are known to have been exposed to various concentrations of TCE since 1980 and of ammonium perchlorate since 1997. It is assumed that perchlorate contamination in the wells was present as early as 1980. The numbers of observed cancers by site and age for the period January 1, 1988, through December 31, 1998, were compared with expected numbers for the same period. Residence at time of diagnosis was used to identify eligible cases. A total of 3,098 cancers occurred among residents during the period. Expected numbers were calculated by applying the average annual age, sex, and race-ethnicity-specific incidence rates within the four-county region for 1988-1992 to the population size of the 13 census tracts reported in the 1990 census and extrapolating population growth through 1998 without accounting for changes in age, sex, or racial or ethnic distributions. Analyses were done for all cancers combined, by specific sites, and separately for children younger than 15 years old.

Standardized incidence ratios (SIRs) were not significantly different from 1.0 (when a conservative 99% confidence interval was used to account for multiple statistical tests) for all cancers combined (SIR, 0.97; 99% CI, 0.93-1.02) or for any specific cancer site, except for colon and rectum (SIR, 0.86; 99% CI, 0.74-0.99) and lung and bronchus (SIR, 0.71; 99% CI, 0.61-0.81), which were lower than expected, and melanoma of the skin (SIR, 1.42; 99% CI, 1.13-1.77) and cancer of the uterine corpus (SIR, 1.35; 99% CI, 1.06-1.70), which were higher than expected. There was neither an excess nor a deficit of thyroid cancer (SIR, 1.00; 99% CI, 0.63-1.47); this estimate was based on 40 observed cases. The types of thyroid cancer were not specified. No cancers were significantly increased in the analysis confined to children, in which all cancers combined, brain tumors, the leukemias, and thyroid cancer were individually examined. The authors speculated that the lower numbers of cancers of some sites were due to lower cigarette-smoking and greater use of screening programs in the community, which is relatively affluent. The higher incidence of uterine cancer was thought to reflect a higher prevalence of postmenopausal estrogen therapy, and it was postulated that the higher incidence of melanoma was a result of failure to adjust completely for the risk associated with fair skin and of the increased use of health care, which led to earlier diagnosis of low-grade melanomas.

No excess of thyroid cancer was observed in the residents, whose drinking water contained measurable TCE and ammonium perchlorate (Morgan and Cassady 2002). The duration of perchlorate contamination of some wells in Redlands before the onset of cancer may have been as short as 8 years or as long as 18 years if contamination began as early as 1980. The duration is uncertain, however, because perchlorate was not measured in well water until 1997. Thus, exposure duration may have been too short to detect outcomes that have long latency, such as thyroid cancer. Expected numbers were derived from the four-county region as a whole, which includes the exposed community, not from a "nonexposed" area. That could result in an underestimate of the SIR. The proportion of the total registry derived from Redlands is not reported. Finally, the authors were unable to adjust for factors that might have confounded the analysis of drinking-water contaminants and cancer in the community.

STUDIES IN NEONATES, CHILDREN, AND PREGNANT WOMEN

Ecologic Studies Based on Newborn Screening Data

Congenital Hypothyroidism

A primary concern regarding exposure to perchlorate in drinking water is its potential effect on the developing fetus and the possibility of inducing or contributing to hypothyroidism in the newborn period. One of the first ecologic studies of perchlorate exposure and thyroid disease in newborns was done by Lamm and Doemland in 1999. The number of observed cases of congenital hypothyroidism in six counties in California and one county in Nevada (Clark County) was compared with that expected on the basis of statewide rates. All cases for 1996-1997 were identified through statewide mandatory neonatal blood screening programs. The criteria used to define congenital hypothyroidism were not stated. Nearly all the water supply for Clark County, which includes Las Vegas, comes from Lake Mead, which is known to be contaminated with perchlorate. Perchlorate in the county's water supply was measured at 4-16 ppb (4-16 μg/L). In California, perchlorate in counties' water supplies is variable, and exposure is intermittent. Nevertheless, the six counties in California were assumed to be "exposed" because they receive water from the Colorado River, in which perchlorate had been detected at 5-8 ppb (5-8 μg/L). Among the nearly 700,000 births in the 2-year period, 249 cases of congenital hypothyroidism were identified, compared with 243 expected, adjusted for Hispanic ethnicity (SIR, 1.0; 95% CI, 0.90-1.16). There were no significant differences between the number of observed cases of congenital hypothyroidism and the number expected for any individual county or for all counties combined. In this study, the expected number was not based on the rates in nonexposed counties but rather on the rate in the entire state, which includes the exposed counties and potentially results in an underestimate of the SIR. Births in the exposed counties amounted to about 60% of all births in the two states. It is implied although not explicit that state-specific rates were used in the individual analyses of Nevada and California data. The expected concentrations were adjusted for county differences in the percentage of newborns of Hispanic ethnicity, but other potential confounding variables, such as birthweight, were not considered.

Thyroid Hormone and TSH Production

A series of ecologic studies examined differences in thyroid hormone and TSH production in newborns in Las Vegas (mean monthly concentrations of perchlorate in drinking water, undetectable to 15 ppb [15 μg/L]) and Reno, Nevada (no detectable perchlorate). The first study compared mean T_4 concentrations—obtained as part of the Nevada newborn blood screening program—in infants born in Las Vegas (n = 17,308) with concentrations in those born in Reno (n = 5,882) over a 15-month period, from April 1998-June 1999 (Z. Li et al. 2000). The infants included had birthweights of 2,500-4,500 g, had blood samples taken within 4 days, and had not been admitted to a neonatal intensive care unit. During that time, Las Vegas drinking water had perchlorate at 9-15 ppb (9-15 μg/L) for 7 months (referred to as period A, April-June 1998 and March-June 1999) and no detectable perchlorate for 8 months (period B, July 1998-February 1999). T_4 concentrations were measured in screening blood specimens taken within the first 4 days of life and in samples taken at infants' first pediatric visits within 60 days of birth. All T_4 analyses were made for the state of Nevada by the Oregon State Public Health Laboratory. The estimated cumulative exposure to perchlorate in drinking water during pregnancy was 0.9-4.2 mg. When stratified by study period (that is, A or B), there was no significant difference in mean T_4 concentrations between infants born in Las Vegas and those born in Reno. In both cities, T_4 values were significantly higher in period B. After adjustment for infants' sex, age at sample collection, and birthweight, there was no significant difference between the two cities in mean newborn T_4 concentrations. There were also no temporal differences between the cities in the relation of mean T_4 concentration and the presence or absence of detectable perchlorate in Las Vegas water. There were no differences between cities in the prevalence of low neonatal T_4 values (at or below the 10th percentile). Postneonatal T_4 concentrations obtained within 60 days of birth also did not differ. Restriction of the study to newborns of normal birthweight (that is, excluding babies under 2,500 g and those over 4,500 g) may have reduced the study's ability to detect differences in mean T_4 concentrations between Las Vegas and Reno newborns. Infants with congenital hypothyroidism have a higher average birthweight, so some cases may have been missed by excluding newborns with birthweights over 4,500 g. Low-birthweight and preterm infants, who were also excluded, are likely to be most vulnerable to the effects of iodide deficiency.

At the time of those studies, Nevada used a two-stage screening procedure for thyroid function in which newborns who had T_4 concentrations below the 10th percentile had a follow-up TSH blood test. Data on TSH concentrations, excluding those obtained in the first day of life, were compared in newborns in Las Vegas (n = 407) and Reno (n = 133) with birthweights of 2,500-4,500 g in the period December 1998-October 1999 (F.X. Li et al. 2000). Crude mean concentrations of TSH did not differ significantly between Las Vegas newborns (11.5 μU/mL) and Reno newborns (12.5 μU/mL). Mean log TSH values also did not vary significantly by city after adjustment for age at specimen collection (2-7 days vs 8-30 days of life) and sex. TSH concentrations were significantly higher in the first age interval than in the second and higher in male than in female infants. Graphic presentation of the trends in mean monthly TSH values for those 2-7 days old by city and mean monthly perchlorate concentrations in Las Vegas water indicated no important differences between newborn TSH values in Las Vegas and Reno and no consistent temporal variations in perchlorate and TSH concentrations in Las Vegas infants. Exclusion of infants with low or high birthweights reduced the potential for confounding of the city comparisons by birthweight. Because a two-stage process was used, only infants with low T_4 values were tested for TSH, so the distributions of TSH concentrations in the two cities might be expected to be similar inasmuch as only one part of the distribution was examined.

Brechner et al. (2000) compared median TSH concentrations, obtained as part of the Arizona newborn blood screening program from October 1994 to December 1997, in newborns in Yuma, which receives all its drinking water from a perchlorate-contaminated source, the Colorado River, and Flagstaff, which receives none of its drinking water from the Colorado River. In a two-stage screening program, infants who had the lowest 10% of T_4 values were retested for TSH. TSH concentrations were determined in about 17% of the 7,599 newborns in Yuma, and in about 15% of the 3,539 newborns in Flagstaff. All assays were done in the Arizona Department of Health Services Laboratory. Perchlorate concentrations in community water were not available for the period of the study. However, 1999 measurements indicated perchlorate at 6 ppb (6 μg/L) in raw and finished water in Yuma and nondetectable concentrations in Flagstaff, and it was assumed that those findings applied to the earlier study period. Median TSH concentrations were statistically significantly higher in Yuma newborns than in Flagstaff newborns (19.9 mU/L vs 13.4 mU/L). A higher percentage of samples were taken before the third day of life from Yuma newborns (69% for non-Hispanic and 71% for Hispanic infants) than from Flagstaff newborns (35% for non-Hispanic and 43% for Hispanic infants). After adjustment for age at sample collection and race or ethnicity, the mean log-transformed TSH values were still statistically significantly higher in Yuma than in Flagstaff ($p = 0.017$). The actual adjusted concentrations for each city were not reported. Neonatal T_4 values did not differ significantly between Yuma and Flagstaff after adjustment for race or ethnicity. However, follow-up testing of TSH is done only in infants with the lowest 10% of T_4 concentrations, so the absence of differences in T_4 concentrations between the cities is not especially informative inasmuch as only the lowest part of the entire distribution was compared. In the Brechner et al. (2000) ecologic study, perchlorate exposures in infants' mothers were not directly measured; in fact, drinking-water concentrations of perchlorate were not derived from the same period as the newborn screening results. It is not known whether water perchlorate concentrations changed between 1994-1997 and 1999, when they were measured for this study, although some data suggest variations in measured exposures of 4-6 ppb (4-6 μg/L) and nondetectable for Yuma during the relevant period (Lamm 2003). Information on birthweight or gestational age was not available, so those potential confounders could not be addressed in the analysis. It has been noted by others (Lamm 2003) that there are several potentially relevant differences between Yuma and Flagstaff other than water perchlorate content. Flagstaff is about 7,000 ft above sea level, and Yuma is near sea level, at 192 ft (U.S. Geological Survey 2004). Infants born at higher elevations often require supplemental oxygen at birth and are generally of lower birthweight than infants born closer to sea level (Nahum and Stanislaw 2004). Low-birthweight infants are more likely to have lower T_4 concentrations without the magnitude of the TSH surge observed in term infants (Mercado et al. 1988). Thus, it is possible that more of the infants with low T_4 who had a follow-up blood test for TSH in Flagstaff were low-birthweight infants who would not have had as great a TSH surge, keeping the mean and median concentrations in Flagstaff lower than in Yuma.

In a commentary and follow-up analysis of the Brechner et al. (2000) study, Lamm (2003) conducted an alternative analysis in an attempt to account for geographic, ethnic, and medical-care differences between Yuma and Flagstaff that might have influenced the results of the earlier analysis. Newborn blood TSH concentrations in Yuma, Flagstaff, and San Luis and Somerton in September 1994-June 1998 were compared. San Luis and Somerton are in Yuma County, near the city of Yuma, but they do not get their municipal water from the Colorado River and thus are assumed to have no exposure to perchlorate. Their combined population is about half that of the city of Yuma. Data on TSH were obtained from the Arizona State Department of Health. The median TSH concentrations were 20.8 mU/L in Yuma, 21.0 mU/L in San Luis and Somerton, and 14.5 mU/L in Flagstaff. Yuma County, as a whole, had the highest median newborn TSH concentrations in the state. Those results suggest that Yuma County as a whole has relatively high TSH screening concentrations, regardless of whether a community has detectable perchlorate in its municipal water supply. However, socioeconomic, racial, ethnic, and low-birthweight differences between Yuma and San Luis and Somerton were not evaluated or controlled, so their influence on these values cannot be assessed.

An unpublished ecologic study by Schwartz (2001) examined the relation of perchlorate exposure to prevalence of abnormal newborn blood screening concentrations of T_4 and TSH and to presumptive and confirmed cases of congenital hypothyroidism. The study included 507,982 California infants born in 1996 for whom newborn screening results were available.

Newborn screening samples were processed by eight laboratories in the state. Exposure to perchlorate was estimated on the basis of tests, conducted in 1997 or later, of samples from drinking-water sources that contributed to 380 public water systems in the state. Testing for perchlorate in California was initiated in systems that were suspected of being contaminated. The 380 systems tested first served about half the state's population. As of 2001, 43% of the water sources of the 380 systems had been tested for perchlorate. The remaining, nontested water systems in the state were assumed to be negative for perchlorate exposure for the purpose of this study. Testing of 820 previously untested systems after the study found that 786 (96%) were negative for perchlorate (J. Schwartz, Impact Assessment, Inc., personal communication, April 9, 2004). Estimates of perchlorate exposure attempted to account for the mixing of water sources, and it was assumed that perchlorate concentrations measured after 1996 reflected those in 1996. Concentrations were averaged across samples from each water source, and average exposures were assigned to each ZIP code. Infants' ZIP codes were based on the mother's residence at time of delivery. Because of the lack of precision of measurements, perchlorate exposure was categorized as none (n = 251,026), low (1-2 ppb [1-2 μg/L]; n = 125,373), medium (3-12 ppb [3-12 μg/L]; n = 129,661), or high (at least 13 ppb [13 μg/L]; n = 1,922).

The unadjusted mean concentrations of T_4 (μg/dL) by perchlorate exposure category were as follows: none, 17.09; low, 16.21; medium, 16.06, and high, 15.05. The crude mean concentrations of TSH (μU/mL) in the same groups were 7.6, 7.6, 7.7, and 7.9. After statistical adjustment for infant sex, ethnicity, multiple birth, birthweight, and age at time of specimen collection, perchlorate category was significantly associated with progressive decrements in mean T_4 concentrations (−0.97, −1.12, and −1.82 μg/dL in the low, medium, and high exposure categories, respectively) and with small but statistically significant increases in ln TSH values (0.029, 0.030, and 0.13 μU/mL). Absolute concentrations for multivariate-adjusted TSH by perchlorate category were not stated. The reductions in T_4 concentrations associated with low birthweight and with early age at specimen collection (7-18 hr after birth) were about 4-7 times those in the highest perchlorate category. Perchlorate category was significantly associated with the presumptive diagnosis of congenital hypothyroidism but not in the predicted direction. Compared with infants living in ZIP codes that had no measurable perchlorate in the water, the odds of presumptive congenital hypothyroidism by category of exposure were 1.15 (95% CI, 1.12-1.17), 1.07 (95% CI, 1.05-1.10) and 1.05 (95% CI, 0.91-1.21) in the low, medium, and high exposure categories, respectively. Because of the small number of confirmed cases of congenital hypothyroidism, perchlorate exposure was dichotomized as "none" vs "any." On the basis of that categorization, there was no significant association between diagnosed congenital hypothyroidism and living in a ZIP code that had measurable perchlorate in the drinking water.

The study of Schwartz was based on a large number of newborns and thus was able to detect small differences in blood screening concentrations of T_4 and TSH among exposure groups if they existed. The clinical importance of the decreases in T_4 with increasing perchlorate is difficult to determine, because the decreases were accompanied by very small changes in TSH. No significant association was observed between perchlorate and confirmed congenital hypothyroidism when exposure was dichotomized. Although important confounders were considered in the analyses, the potential for misclassification of perchlorate exposure remains. No measurement of perchlorate had been done in a substantial percentage of water sources that were used to characterize each water system; it was assumed that later measurement reflected 1996 measurements and that nontested systems were negative, and, as in other ecologic studies, there was no measurement of perchlorate exposure in the water to which individual infants were exposed. Nondifferential misclassification of exposure tends to bias associations toward the null value, but only for dichotomous exposure variables. When exposures are categorized at more than two levels, nondifferential misclassification can introduce biases both toward and away from the null in the same dataset (Rothman and Greenland 1998). In addition, the investigator was unable to control for laboratory variation in thyroid hormone and TSH measurements. Because eight laboratories in California are responsible for conducting newborn screening tests, small variations in results among geographic areas served by different laboratories may have contributed to the observed differences.

Kelsh et al. (2003) examined the frequency of congenital hypothyroidism and of high blood TSH concentrations among newborns in 1983-1997 whose mothers resided in one of two California communities and who were screened as part of the California Newborn Screening Program. Measurements of perchlorate in drinking water were made by the California Department of Health Services Drinking Water Program, which began testing in 1997. The "exposed" community included 13 census tracts in Redlands and Mentone serviced by the Redlands Municipal Water District in which perchlorate had been detected in the water system at up to 9 ppb (9 μg/L), with a calculated mean concentration below 1 ppb (1 μg/L). The comparison communities were San Bernardino and Riverside counties, which were adjacent to Redlands and used the same newborn screening laboratories but in which perchlorate had not been detected in the water supply. Because assessment of thyroid function was ascertained in the immediate newborn period, the relevant exposure was assumed to be fetal. Exposure data obtained in 1997 were assumed to apply to the entire study period. Cases were infants in whom congenital hypothyroidism was diagnosed or whose screening concentration of TSH was "elevated," usually defined as over 25 μU/mL but sometimes over 16 μU/mL. During the study period, the California Newborn Screening Program measured TSH only among infants who first screened in the lower 5% of T_4 concentrations. Thus, TSH concentrations are available only for infants who also had low T_4 concentrations. Potentially confounding variables included in the analyses were age at specimen collection (in hours since birth, categorized as less than 6, 6 to less than 12, 12 to less than 24, and 24 and higher), sex, race or ethnicity, birthweight, multiple birth status (yes or no), and calendar year of birth.

About two-thirds of the newborns in both "exposed" and nonexposed communities had their TSH screening blood specimens collected at 18 hr or more after birth (Kelsh et al. 2003). A total of two cases of congenital hypothyroidism were found among the 15,348 Redlands births, compared with the expected 4.66 cases based on data from San Bernardino and Riverside births. The standardized prevalence ratio (SPR) for congenital hypothyroidism in Redlands was 0.45 (95% CI, 0.06-1.64), adjusted for infant's sex, ethnicity, birthweight, and birth year. Multivariate analysis found no statistically significant relation between residence in Redlands and the odds of "elevated" TSH among newborns whose specimens were collected 18 hr or more after birth (odds ratio [OR], 0.72; 95% CI, 0.28-1.54). The analysis was based on six cases of "elevated" TSH in the Redlands newborns. Additional analyses, specifically excluding from the control communities infants who were born in areas with any measurable perchlorate or infants in communities that receive water from the Colorado River, did not alter the results in any meaningful way.

As an ecologic study, the Kelsh et al. (2003) study is limited in that exposure of individual mothers is unknown. In addition, exposure data from a single year were used to characterize exposures over the entire 15 years of the study. Whether residence in areas with measurable perchlorate was associated with adverse thyroid outcomes in infants of low birthweight was not specifically addressed. Birthweight was analyzed as a continuous variable in the multivariate analysis, in spite of the reported observation of a U-shaped relation between birthweight and TSH concentrations.

Buffler et al. (2004) extended the Kelsh et al. (2003) study by analyzing data from the California Newborn Screening program for 1998. They included all births in the state to mothers residing in communities in which the drinking water had been tested for perchlorate in 1997-1998. "Exposed" communities were those in which the mean perchlorate concentration in drinking water was over 5 ppb (5 μg/L), and "nonexposed" communities were those in which perchlorate was not detectable or the average concentration was 5 ppb (5 μg/L) or lower. Data on congenital hypothyroidism and high blood TSH (generally defined as a screening value over 25.0 μU/dL) were compared. In 1998, the two-tiered TSH sampling strategy was discontinued, and all samples collected at the age of 24 hours or later were included. SPRs for congenital hypothyroidism were calculated as the ratio of the observed number of cases in the "exposed" group to the number expected on the basis of rates in the "nonexposed" group, standardized for ethnicity, sex, and birthweight. The odds of congenital hypothyroidism and of high TSH values (categorized as "high" or "normal") in "exposed" compared with "nonexposed" newborns was estimated with logistic regression methods, with adjustment for infants' sex, ethnic status, birthweight (continuous for TSH analysis, categoric for primary hypothyroidism analysis), year of birth, and age at blood collection (for TSH). The authors reported no statistically significant relation between exposure to perchlorate over 5 ppb (5 μg/L) and the prevalence of or odds of congenital hypothyroidism or increased TSH values (Buffler et al. 2004). Their analysis addressed some of the limitations noted in the earlier study of Kelsh et al. (2003). Specifically, the timing of the drinking-water assessments was concordant with the birth cohort, and because the data covered the entire state, the observed numbers of cases of congenital hypothyroidism (15) and of increased TSH level (147) in the "exposed" areas were considerably higher than in the prior study. However, exposure assessment on an individual level was still lacking.

Thyroid Hormone and TSH Production in Children

Crump et al. (2000) compared thyroid hormone and TSH production and the frequency of thyroid disease in newborns and in school-age children in three Chilean cities with different perchlorate concentrations in their municipal water supplies. Perchlorate was measured in drinking-water samples in each city drawn from potable-water faucets in schools, homes of the children, and public buildings near the schools. This is the only study in children that was based on outcome measures in individual children and on perchlorate measurements in water from taps accessible to the children for drinking water. The analysis of data from newborn screening is a purely ecologic design. The three cities were Antofagasta (no detectable perchlorate), Chanaral (5-7 ppb [5-7 μg/L]), and Taltal (100-120 ppb [100-120 μg/L]). The cities are west of the Andes Mountains in the Atacama Desert of northern Chile, one of the most arid regions on Earth. The region receives measurable rainfall as infrequently as once in 5-20 years, so there is no farming. Water is supplied to Antofagasta from a pipeline on the western margins of the Andes, an area that contains large amounts of naturally occurring perchlorate. Potable water for all other cities in the area comes from ground water in wells in alluvial basins. All water samples were analyzed in the same laboratory, and technicians did not know the city from which the samples came.

For the study of school-age children, first- and second-graders 6-8 years old (50-60 children per city) were recruited from one or two public schools in each city in 1999. They were of similar ethnicity and socioeconomic status. More than 90% of their parents agreed to their participation; this involved completion of a questionnaire by parents and collection of blood and urine specimens from the children. The questionnaire included items related to family history of thyroid disease and residential history. Blood samples were drawn from the 162 participating children between 9 a.m. and 3 p.m. Studies included serum TSH, T_4, free T_4, and T_3 concentrations and tests of liver and kidney function. A first-void, spot urine sample was obtained for measurement of urinary iodide and creatinine. Children were also examined for evidence of goiter by one of three endocrinologists who did not know the perchlorate concentrations in the study cities. Group analyses were performed for the 127 children who had lived their entire lives, from conception to testing, in the same city. Neonatal thyroid screening has been mandatory in Chile since 1992, and all samples for the entire country are processed in a single facility. For the present study, newborn samples were obtained from the 9,784 children born from February 1996 to January 1999 in the three cities.

Among children who were life-long residents in their city, there were no statistically significant differences in serum TSH levels among the three cities after adjustment for age, sex, and urinary iodide concentration (Crump et al. 2000). Serum free T_4 levels were significantly higher in children in Chanaral and Taltal than Antofagasta after adjustment for age, sex, and urinary iodide levels; the difference was opposite the direction predicted on the basis of competitive perchlorate inhibition of iodide uptake. There were no significant differences among the cities in the prevalence of goiter among the schoolchildren (22%, 20%, and 26% in Antofagasta, Chanaral, and Taltal, respectively). Compared with Antofagasta, however, the multivariate-adjusted odds of having a relative with a history of goiter, hypothyroidism, or subtotal thyroidectomy was nearly 5 times as high in Taltal (OR, 4.97; 95% CI, 1.29-19.17) and was not significantly different in Chanaral (OR, 1.04; 95% CI, 0.21-5.09). Family history of goiter was reported on the study questionnaire and included parents, siblings, grandparents, great-grandparents, aunts, uncles, and cousins. Crump et al. (2000) suggest that the high prevalence of a family history of goiter, which was seen only in Taltal, may have resulted from a combination of high perchlorate and low iodide intake before the introduction of iodized salt into the region in 1982. The country has since been identified by the Iodine Deficiency Disease Prevalence and Control Program as an area of iodide excess, with a population median urinary iodide excretion of 54 μg/dL (International Council for the Control of Iodine Deficiency Disease 2001, 2002).

Multivariate-adjusted comparisons for serum T_3 concentrations were not presented, but univariate analyses indicated no differences in mean values among the three cities for all children (Crump et al. 2000). Mean urinary iodide levels in the 6- to 8-year-olds did not differ significantly: 75.6 ± 40.4 μg/dL in Antofagasta, 61.4 ± 35.7 μg/dL in Chanaral, and 76.6 ± 47.4 μg/dL in Taltal. No statistically significant differences were observed among children from the three cities in measures of bone marrow, liver, or kidney function.

After adjusting for sex and age of testing (in days), newborns in Taltal had statistically significantly lower log serum TSH concentrations; mean log TSH levels did not vary significantly between newborns in the other two cities (Crump et al. 2000). The lower mean in Taltal is, again, opposite of what would be expected in association with increased exposure to perchlorate. During the period of observation, seven presumptive cases of congenital hypothyroidism (TSH, at least 25 μU/mL) were identified, all in infants born in Antofagasta.

The committee thinks that the study of Crump et al. (2000) had important strengths. Although individual exposures were not assessed, it is one of the few studies that measured perchlorate in drinking water in samples taken directly from the environment of the children studied, such as homes and schools. In addition, it was possible to compare assessments of thyroid function, other end points, and potential risk factors obtained in a systematic manner from all participants and adjusted for a number of important covariates, and participation was high. All laboratory assessments were done at the same facility, and assessments of thyroid status and other measures were done by observers unaware of the perchlorate exposure of the children. All newborn screening tests in Chile are done at a single laboratory.

Numerous critiques of the Crump et al. (2000) study have been done either directly by the U.S. Environmental Protection Agency (EPA) in its risk-assessment document or by others at the request of EPA (Park 2001; Marcus 2003). Reanalysis by EPA was also done in 2002 with only the published data. The authors have responded to each critique. With respect to the analyses done among school-age children, the study was criticized for failing to adjust for differences among the three cities in ethnicity and socioeconomic status. Ethnicity has been shown in some studies to be related to risk of congenital hypothyroidism or to thyroid hormone concentrations, at least in newborns (Lamm and Doemland 1999; Brechner et al. 2000; Schwartz 2001; Kelsh et al. 2003). The authors provided data to support their contention that there are no important differences among the three cities in ethnicity, socioeconomic status, or access to medical care (Gibbs 2003a). Many of the data supplied in the Gibbs (2003a) letter come from a study of pregnant women in the three cities, which is described below.

The analyses of newborn screening data have been criticized on the basis of failure to adjust for differences among the cities in iodide intake, ethnicity, and birthweight. No other studies have included adjustment for maternal iodide intake, so this concern is not peculiar to the study in Chile. In the analysis of data from school-age children, there were no differences in urinary iodide excretion between the lowest- and highest-exposure cities; urinary iodide in Chanaral (the middle-exposure city) was about 15 μg/dL lower than those in Antofagasta and Taltal. Similar findings are reported in unpublished preliminary data from the study of pregnant women. The committee does not think that failure to adjust for maternal iodide intake can explain the significantly *lower* TSH concentrations in Taltal newborns. There are no differences in ethnicity, so it is not necessary to adjust for this variable. Finally, comparisons of newborn thyroid hormone screening concentrations were criticized because of failure to adjust for birthweight. Information was not provided on the distributions of birthweights or gestational ages in the three cities, so it is not possible to assess whether the newborn screening results might have been influenced by differences in them.

The study was also criticized because of differences in the median postnatal day on which newborn samples were tested for serum TSH. The committee considers this issue adequately addressed by Crump et al. (2000), in that they compared TSH concentrations by single days of age (Table 8, p. 609), and age at sample collection was included in the multivariate analysis of TSH (Table 9, p. 610).

The possibility was raised that differences among the three cities in ambient indoor and outdoor temperatures may have affected measures of thyroid hormone in newborns. The authors provide data to show that the climate is very similar in the three areas and that, in any case, no differences would be expected in the delivery suites in the three locations (Gibbs 2003a). The committee considers this criticism adequately addressed by the authors.

The Marcus (2003) critique also points to high coefficients of variation (CVs) in the endocrine measures of the study by Crump et al. (2000), citing CVs (%) of 45-54% for serum TSH and 50-64% for urinary iodide. Marcus (2003) suggests that this variability is evidence of poor laboratory measurement techniques and an absence of quality-control procedures. In response, the authors provide data comparing the CVs for measurements in their study with those of other published work (Gibbs and Crump 2003). Those data represent interassay variation rather than interindividual variation and indicate no differences in precision of measures between Crump et al. (2000) and reports from other laboratories.

Many of the issues raised in regard to the study by Crump et al. (2000) apply equally to other ecologic studies and are not peculiar to the study done in Chile. They include the possibility of "uncontrolled confounders," which can be present in any epidemiologic study, and questions regarding the characterization of exposure.

The major criticisms of the study of Crump et al. (2000) are related to the high dietary iodide intake in the populations and to the high prevalence of goiter in the children themselves (as assessed by physical examination) and their family members (based on mothers' reports). EPA considered the high prevalence of goiter to be an indication of thyroid abnormalities in the population and further evidence of the unsuitability of the data for inference to the U.S. experience. The mean urinary iodide excretion in children in the three Chilean cities (61-77 μg/dL) is about 3 times as high as that in U.S. children 6-9 years old in 1988-1994 (geometric means, 25.2 and 20.3 μg/dL in boy and girls, respectively) (Crump et al. 2000; Soldin et al. 2003). The prevalence of goiter at which iodide deficiency is considered to be a public-health problem was defined by the World Health Organization (WHO) in 1994 as 5% (WHO 1994). The proportions of lifelong resident, 6- to 8-year-old children who had clinically diagnosed goiter were 22%, 20%, and 26% in Antofagasta, Chanaral, and Taltal, respectively, exceeding the WHO value of 5% in spite of the apparently high iodide intake. Goiter in the children was assessed by palpation, and the presence of goiter was based on the 1960 WHO criteria, which, although more valid than the revised, 1994 WHO criteria, has a specificity of only about 76% in a high-prevalence area (Peterson et al. 2000). Basing the diagnosis of goiter solely on the results of palpation has been shown to have good reproducibility within observers but not between observers and to have poor specificity compared with ultrasonography, resulting in a high number of false positives (Peterson et al. 2000). Therefore, the committee did not consider the higher than expected prevalence of goiter to be grounds for dismissing the results of the Chilean study.

It is suggested in EPA's critiques that the high dietary iodide intake makes this geographic area unsuitable for studies of the thyroid effects of perchlorate exposure. The contention is that the iodide intake is so high that substantial competition with perchlorate cannot be detected. Because EPA did not supply data to support that contention, the committee performed calculations on the effect of basal serum iodide concentrations in humans on the inhibition of iodide uptake by the thyroid sodium (Na^+)/iodide (I^-) symporter (NIS) in the presence of perchlorate. The Michaelis-Menten competitive-inhibition equation was used for the calculations; the assumptions used in each model are included in Appendix D. A second set of calculations produced a series of curves of perchlorate's inhibition of iodide uptake across a 10^5-fold range of basal serum iodide concentrations (Appendix D). On the basis of those models, serum iodide concentrations in humans over a 10-fold range (1.5-15 μg/dL) are not likely to have a profound influence on the ability to detect a perchlorate effect on the thyroid NIS. In addition, a basal serum iodide concentration over 100 μg/dL would be required to shift the dose-response curve for perchlorate's inhibition of iodide uptake; this suggests that serum iodide concentrations within 0-100 μg/dL would be equally sensitive to perchlorate's effects. On the basis of the calculations, the committee concludes that iodide intake in the population in question does not interfere with the study's ability to detect effects of perchlorate exposure at up to 120 ppb (120 μg/L) on iodide uptake, if effects are present. It should also be noted that in spite of the contention that dietary iodide is too high in the population to allow the detection of a perchlorate effect, an effect on serum free T_4 concentrations was observed (albeit inverse). The mean serum T_4 value was highest in Taltal, the city with the highest exposure to perchlorate.

In addition to those calculations, the committee compared serum concentrations of TSH and thyroid hormones reported by Crump et al. (2000)

in children 6-8 years old with those children of similar ages in the United States during a comparable period (Zurakowski et al. 1999) as another approach to assessing whether data from the Chilean study might be useful for evaluating the U.S. experience. With the exception of serum T_3, which is known to vary more depending on the assay used, the mean serum thyroid hormone concentrations were not markedly different between U.S. and Chilean children. The mean TSH concentrations were somewhat higher in Chile but were not accompanied by decreases in serum T_4 concentrations. Those results suggest that mean serum thyroid hormone concentrations in Chilean children were similar during the period to those in U.S. children of the same age. On the basis of the iodide-inhibition analyses, the additional comparisons, and a review of information on urinary iodide excretion, the committee concluded that the data from Chile could be considered in the evaluation of the U.S. experience with perchlorate in drinking water.

Thyroid Hormone and TSH Production in Pregnant Women and Their Newborns

A second study among pregnant women and their newborns in each of the three cities in Chile examined the relation between perchlorate exposure during pregnancy, indexes of maternal thyroid function, and thyroid hormone concentrations in the newborns (Gibbs 2003b). Serum T_3, free T_4, TSH, thyroid peroxidase antibodies, antithyroglobulin antibodies, and thyroglobulin and urinary iodide were measured in mothers during the first and third trimesters and after birth. Maternal blood specimens from the first trimester, maternal urine specimens from the first and third trimesters and after birth, umbilical cord blood, and samples of breast milk were tested for perchlorate. Water samples from the pregnant women's homes were analyzed for perchlorate to provide individual home-exposure estimates.

Preliminary unpublished results on maternal urinary iodide, serum T_3, free T_4, thyroglobulin, and TSH concentrations at the first and second prenatal visits, cord blood thyroid hormones, and perchlorate in breast milk were presented at a public meeting in May 2004 (Gibbs 2004a). Analyses of accrued data were provided to the committee in a written report in August 2004 (Gibbs 2004b) and are presented here. Mean urinary iodide in mothers at the first prenatal visit was significantly different among the three cities, with mean levels being lowest in Taltal (323 μg/g creatinine in 62 samples compared with 407 μg/g creatinine in 61 samples in Antofagasta, and 363 μg/g creatinine in 39 samples in Chanaral). Maternal mean serum free T_4 and TSH concentrations from the first prenatal visit were similar among the three cities, although mothers in Chanaral had the highest mean TSH values (2.81 μU/mL vs 2.63 μU/mL in Antofagasta and 2.61 μU/mL in Taltal). Mean serum T_3 was significantly different in first-trimester samples. Mean T_3 was similar in mothers in Antofagasta and Taltal (183 ng/dL and 187 ng/dL, respectively), but it was higher in Chanaral (207 ng/dL). Mothers identified at the first prenatal visit as hypothyroid (serum TSH, at least 4.5 μU/dL) were treated. Five such mothers were identified in each of the three cities. At the second prenatal visit, serum T_3 values were significantly different, with the lowest mean in Taltal (173 ng/dL), but they were not adjusted for week of gestation, which was longest at the time of sampling in Taltal (33.2 weeks). Mean serum free T_4 was also significantly different among the cities. Serum free T_4 values were similar in Chanaral and Taltal (0.82 ng/dL and 0.83 ng/dL, respectively), but higher in Antofagasta (0.86 ng/dL), again, no adjustment was made for differences in length of gestation. Median iodide concentrations in samples of breast milk from mothers were 36.9 μg/dL in Antofagasta (16 samples), 29.5 μg/dL in Chanaral (16 samples), and 38.4 μg/dL in Taltal (25 samples). Median perchlorate concentrations in breast milk were less than 0.5, 19.3, and 103.8 ppb (μg/L) respectively (Gibbs 2004b).

Only mean T_3 was significantly different in cord blood samples; the value was lowest in Chanaral at 73 ng/dL compared with 79 ng/dL in Antofagasta and 82 ng/dL in Taltal. In data presented in May but not updated in August, perchlorate was detectable in cord blood specimens from infants born in Taltal, but not in Chanaral. It should be noted that no analyses have been adjusted by gestational age or birthweight, so the results presented for newborns are preliminary. The study had not been published at the time of the committee's deliberations; therefore, it did not use the results in formulating its conclusions.

Ecologic Studies of Neurodevelopment in Children

There are sparse data concerning the relation between perchlorate exposure and neurodevelopmental disorders in children. One such study is that of attention-deficit-hyperactivity disorder (ADHD) and autism by Chang et al. (2003), which examined the association between residence in Nevada communities with and without detectable perchlorate in the public water supply and diagnoses of either ADHD or autism in children less than 18 years old who were recipients of Medicaid. Scores on tests of 4th grade performance were also compared with national values for 1998-1999 and 2001. The three comparison groups were in Clark County, which includes Las Vegas and in which the public water supply has perchlorate ranging from undetectable to 23.8 ppb (23.8 µg/L) (median, 10.5 ppb [10.5 µg/L]) as measured in 1997-2001; Washoe County, which includes Reno and has no detectable perchlorate in the public water supply; and the remainder of Nevada, was considered a "rural" control for Las Vegas. The rural areas have no detectable perchlorate in public water supplies. Data from the service records of the Nevada Medicaid Program for calendar years 1996-2000 were used to ascertain the total number of Medicaid recipients less than 18 years old for each year and the number of cases with a first, underlying diagnosis of ADHD or autism. The midpoint Medicaid population size for that period and the average number of cases per year in each of the three geographically defined groups were used to calculate the average annual proportion of Medicaid recipients under 18 years old with ADHD or autism. For ADHD, the annual numbers of cases per 1,000 Medicaid recipients under 18 years old were 17, 28, and 29 for Clark, Washoe, and the rest of the state, respectively; for autism, the annual numbers of cases per 1,000 were 1.4, 2.8, and 1.2.

Thus, Clark County, the only area exposed to perchlorate in drinking water, did not have a higher proportion of ADHD or autism than non-exposed communities. Similarly, 4th grade proficiency scores, expressed as national percentile ranks, did not differ significantly among the three areas. A comparison of 2001 4th grade achievement scores allowed inclusion only of children who had resided in the same community for at least 3 years. The results also showed no important differences with respect to 4th grade achievement among the three exposure groups. Other than the single analysis of the 2001 4th grade achievement scores, the data cannot account for any changes in children's residence. They also do not reflect exposures to perchlorate at a time in neurologic development that is considered biologically relevant, at least for autism, namely early gestation. As an ecologic study, the study measured neither exposure nor outcome in individuals. In addition, the specific criteria used to make individual diagnoses of ADHD or autism are unknown; diagnoses were made by multiple health-care providers and without the use of uniform diagnostic criteria. Thus, the validity of the diagnoses cannot be assessed, and geographic differences could reflect diagnostic practices in local communities rather than true differences in risk. Furthermore, no adjustment was made for potentially confounding factors, such as age, sex, race, and ethnicity. Finally, the end points studied are not the most sensitive indicators of the kinds of neurodevelopmental outcomes that might be predicted on the basis of prior studies of the effects of hypothyroidism on the developing nervous system. The indicators include subtle impairments in cognitive and motor function, such as those observed in children who have untreated or inadequately treated congenital hypothyroidism. In addition, neither autism nor autistic-spectrum disorder has been observed previously in association with thyroid hormone deficiencies.

SUMMARY

Limitations of Existing Data

Epidemiologic studies have examined possible associations between environmental exposure to perchlorate in drinking water at about 4-120 ppb (4-120 μg/L) and abnormalities of thyroid hormone and TSH production in newborns, as well as thyroid diseases (such as congenital hypothyroidism and goiter) and cancer in children and adults (Lamm and Doemland 1999; Brechner et al. 2000; Crump et al. 2000; F.X. Li et al. 2000; Z. Li et al. 2000; Schwartz 2001; Morgan and Cassady 2002; Kelsh et al. 2003; Lamm 2003; Buffler et al. 2004). Occupational studies of respiratory exposures up to 0.5 mg/kg perchlorate per day and abnormalities of thyroid hormone and TSH production in adult workers have been conducted (Gibbs et al. 1998; Lamm et al. 1999; Braverman et al. 2004). Only one study has examined a possible relation between perchlorate exposure and adverse neurodevelopmental outcomes (ADHD and autism) in children (Chang et al. 2003). A number of the studies have samples that are too small to detect differences in frequency of outcomes between exposure groups, and adjustment for potentially confounding factors was limited. Nearly all the studies were ecologic, including those in newborns and children, the groups potentially most vulnerable to the effects of perchlorate exposure. Ecologic studies can provide supporting evidence of a possible association but cannot themselves provide definitive evidence regarding cause. Perchlorate exposure of individuals is difficult to measure and was not assessed directly in any of the studies conducted outside the occupational setting. The only study with measures of perchlorate made directly from drinking-water samples taken from faucets potentially used by people who were studied was that done in Chile by Crump et al. (2000).

No studies have examined the relation of perchlorate exposure and adverse outcomes, either in thyroid function or in neurodevelopment, among especially vulnerable groups, such as low-birthweight or preterm infants. In addition, the available studies do not assess the possibility of adverse outcomes associated with perchlorate exposure in infants born to mothers who had inadequate dietary iodide intake. Thus, no direct human data are available regarding a possible interaction between maternal iodide intake and perchlorate exposure.

Although the ecologic design is inherently limited with respect to establishing causality, results of such studies can be informative when combined with other data on the biology of the thyroid gland, experimental studies of the effects of acute exposure to perchlorate, and studies of occupational perchlorate exposure.

Committee Conclusions Drawn from Epidemiologic Data on Specific Health Outcomes

Congenital Hypothyroidism

Ecologic data alone are not sufficient to demonstrate whether or not an association is causal, but they do provide evidence that can be used to evaluate possible associations. Acknowledging that ecologic data alone are inherently limited in drawing causal inferences, the committee concludes that the available epidemiologic evidence is not consistent with a causal association between perchlorate exposure and congenital hypothyroidism as defined by the authors of the studies reviewed here. All studies of this association were negative.

Perturbation of Thyroid Hormone and TSH Production in Newborns

Again given the limitations of ecologic data in inferring causation, the available epidemiologic evidence is not consistent with a causal association between perturbations of thyroid hormone and TSH production in normal newborns (that is, not low-birthweight or preterm) and exposure during gestation to perchlorate in drinking water at up to 120 ppb (120 μg/L). Most studies do not show either significantly lower T_4 concentrations or significantly higher TSH concentrations among infants born in geographic areas in which the water supply has measurable perchlorate.

However, no epidemiologic studies are available on the association between perchlorate exposure and thyroid dysfunction among low-birth-weight or preterm newborns, offspring of mothers who had iodide deficiency during gestation, or offspring of hypothyroid mothers. Those are the groups of greatest concern with respect to potential effects of perchlorate exposure.

Neurodevelopmental Outcomes

The epidemiologic evidence is inadequate to determine whether or not there is an association between perchlorate exposure and adverse neurodevelopmental outcomes in children. The only relevant study used an ecologic design and examined autism and ADHD as end points. Subtler neurodevelopmental outcomes have not been assessed in human populations. Although inclusion of ADHD was considered plausible, the committee questioned the appropriateness of autism as an end point. Autism has not been observed among the spectrum of clinical outcomes in children who had congenital hypothyroidism and were evaluated prospectively (Rovet 1999, 2002, 2003).

Thyroid Diseases and Hypothyroidism in Adults

On the basis of data from studies of chronic occupational exposures to ammonium perchlorate and ecologic investigations in adults, the committee concludes that the epidemiologic evidence is not consistent with a causal association between exposure to perchlorate at the concentrations studied and thyroid diseases in adults. The thyroid diseases and thyroid measures investigated included simple and nonspecified goiter, nontoxic nodular goiter, thyrotoxicosis with or without goiter, acquired hypothyroidism, thyroiditis, and other disorders of the thyroid, and measures of serum TSH and the two thyroid hormones. In occupational studies, perchlorate exposures as high as 0.5 mg/kg per day have not been associated with adverse effects on thyroid function in workers, but small samples in some studies may have reduced the ability to identify important differences, and studies were limited to workers who remained in the workforce.

Thyroid Cancer in Adults

The epidemiologic evidence is insufficient to determine whether there is a causal association between perchlorate exposure and thyroid cancer. Only two studies related to this issue have been done, and both were ecologic. In one study, the number of thyroid-cancer cases was too small to have a reasonable chance of detecting an association if one existed (Li et al. 2001). In the second, larger study (Morgan and Cassady 2002), mixed exposures were present (to perchlorate and TCE). In neither study was it possible to adjust for potential confounding variables. The committee notes, however, that on the basis of its understanding of the biology of human and rodent thyroid tumors, it is unlikely that perchlorate poses a risk of thyroid cancer in humans.

REFERENCES

Braverman, L.E., X. He, S. Pino, M. Cross, B. Magnani, S.H. Lamm, K. Crump, and J. Gibbs. 2004. The effect of perchlorate, thiocyanate, and nitrate on thyroid function in long-term workers exposed to perchlorate. Presentation at the Fourth Meeting of the Committee to Assess the Health Implications of Perchlorate Ingestion, May 24, 2004, Woods Hole, MA.

Brechner, R.J., G.D. Parkhurst, W.O. Humble, M.B. Brown, and W.H. Herman. 2000. Ammonium perchlorate contamination of Colorado River drinking water is associated with abnormal thyroid function in newborns in Arizona. J. Occup. Environ. Med. 42(8):777-782.

Buffler, P.A., M.A. Kelsh, E.C. Lau, C.H. Edinboro, and J.C. Barnard. 2004. Epidemiologic Studies of Primary Congenital Hypothyroidism and Newborn Thyroid Function Among California Residents, Final Report. April 2004, Berkeley, CA.

Chang, S., C. Crothers, S. Lai, and S. Lamm. 2003. Pediatric neurobehavioral diseases in Nevada counties with respect to perchlorate in drinking water: An ecological inquiry. Birth Defects Res. Part A Clin. Mol. Teratol. 67(10):886-892.

Crump, C., P. Michaud, R. Tellez, C. Reyes, G. Gonzalez, E.L. Montgomery, K.S. Crump, G. Lobo, C. Becerra, and J.P. Gibbs. 2000. Does perchlorate in drinking water affect thyroid function in newborns or school-age children? J. Occup. Environ. Med. 42(6):603-612.

Gibbs, J.P., R. Ahmad, K.S. Crump, D.P. Houck, T.S. Leveille, J.E. Findley, and M. Francis. 1998. Evaluation of a population with occupational exposure to airborne ammonium perchlorate for possible acute or chronic effects on thyroid function. J. Occup. Environ. Med. 40(12):1072-1082.

Gibbs, J.P. 2003a. Comments and Some Information About Specific Issues, Rafael Tellez. Letter to R. Johnston, Committee to Assess the Health Implications of Perchlorate Ingestion, from J.P. Gibbs, Kerr-McGee Corp., Oklahoma City, OK. December 16, 2003.

Gibbs, J.P. 2003b. Comments Regarding Previous and Ongoing Studies in Northern Chile. Protocol and Preliminary Data from an Ongoing Study Among Pregnant Women and Newborns in the Same Three Cities as Studied by Crump et al (2000). Letter to E.K. Mantus, National Academy of Sciences, from J.P. Gibbs, Kerr-McGee Corp., Oklahoma City, OK. October 27, 2003.

Gibbs, J.P. 2004a. Preliminary: Chronic Environmental Exposure to Perchlorate in Drinking Water and Thyroid Function During Pregnancy and the Neonatal Period. Presentation at the Fourth Meeting of the Committee to Assess the Health Implications of Perchlorate Ingestion, May 24, 2004, Woods Hole, MA.

Gibbs, J.P. 2004b. Chronic Environmental Exposure to Perchlorate in Drinking Water and Thyroid Function During Pregnancy and the Neonatal Period. August 8, 2004 Update. Letter to R. Johnston, Committee to Assess the Health Implications of Perchlorate Ingestion, from J.P. Gibbs, Kerr-McGee Corp., Oklahoma City, OK. August 7, 2004.

Gibbs, J., and K. Crump. 2003. Analysis of Review of Original Statistical Methods and Alternate Analyses for Ecological Epidemiological Study of Crump et al. (2000) by A.H. Marcus, National Center for Environmental Assessment, in memorandum to A.M. Jarabek, National Center for Environmental Assessment, Washington, DC, September 23, 2003. December 11, 2003.

International Council for the Control of Iodine Deficiency Disease. 2001. The Western Hemisphere Nears Iodine Sufficiency. IDD Newsletter 17(1):2. [Online]. Available: http://www.people.virginia.edu/%7Ejtd/iccidd/newsletter/feb2001.htm [accessed July 8, 2004].

International Council for the Control of Iodine Deficiency Disease. 2002. Iodine Deficiency Disease (IDD) Prevalence and Control Program Data: Chile. [Online]. Available: http://www.people.virginia.edu/~jtd/iccidd/mi/idd_034.htm [accessed July 8, 2004].

Kelsh, M.A., P.A. Buffler, J.J. Daaboul, G.W. Rutherford, E.C. Lau, J.C. Cahill, A.K. Exuzides, A.K. Madl, L.G. Palmer, and F.W. Lorey. 2003. Primary congenital hypothyroidism, newborn thyroid function, and environmental perchlorate exposure among residents of a southern California community. J. Occup. Environ. Med. 45(10):1116-1127.

Lamm, S. H. 2003. Perchlorate exposure does not explain differences in neonatal thyroid function between Yuma and Flagstaff. [Letter]. J. Occup. Environ. Med. 45(11):1131-1132.

Lamm, S.H., and M. Doemland. 1999. Has perchlorate in drinking water increased the rate of congenital hypothyroidism? J. Occup. Environ. Med. 41(5):409-411.

Lamm, S.H., L.E. Braverman, F.X. Li, K. Richman, S. Pino, and G. Howearth. 1999. Thyroid health status of ammonium perchlorate workers: A cross-sectional occupational health study. J. Occup. Environ. Med. 41(4):248-260.

Li, F.X., D.M. Byrd , G.M. Deyhle, D.E. Sesser, M.R.Skeels, S.R.Katkowsky, and S.H. Lamm. 2000. Neonatal thyroid-stimulating hormone level and perchlorate in drinking water. Teratology 62(6):429-431.

Li, Z., F.X. Li, D. Byrd, G.M. Deyhle, D.E. Sesser, M.R. Skeels, and S.H. Lamm. 2000. Neonatal thyroxine level and perchlorate in drinking water. J. Occup. Environ. Med. 42(2):200-205.

Li, F.X., L. Squartsoff, and S.H. Lamm. 2001. Prevalence of thyroid diseases in Nevada counties with respect to perchlorate in drinking water. J. Occup. Environ. Med. 43(7):630-634.

Marcus, A.H. 2003. Review of Original Statistical Methods and Alternate Analyses for Ecological Epidemiology Study of Crump et al. (2000). Memorandum to A.M. Jarabek, National Center for Environmental Assessment, from A.H. Marcus, National Center for Environmental Assessment, Washington, DC. Sept. 23, 2003. [Online]. Available: http://www.epa.gov/ncea/perchlorate/references/documents/crump_memo.pdf [accessed September 30, 2004].

Mercado, M., V.Y. Yu, I. Francis, W. Szymonowicz, and H. Gold. 1988. Thyroid function in very preterm infants. Early Hum. Dev. 16(2-3):131-141.

Morgan, J.W., and R.E. Cassady. 2002. Community cancer assessment in response to long-time exposure to perchlorate and trichloroethylene in drinking water. J. Occup. Environ. Med. 44(7):616-621.

Nahum, G.G., and H. Stanislaw. 2004. Hemoglobin, altitude and birth weight: Does maternal anemia during pregnancy influence fetal growth? J. Reprod. Med. 49(4):297-305.

Park, R.M. 2001. Perchlorate Health Effects in the Epidemiological Literature. Letter to A. Jarabek, National Center for Environmental Assessment, Office of Research and Development, Washington, DC, from R.M. Park, National Institute for Occupational Safety and Health, Cincinnati, OH. [Online]. Available: http://www.epa.gov/ncea/perchlorate/references2/documents/100842.pdf [accessed September 30, 2004].

Peterson, S., A. Sanga, H. Eklof, B. Bunga, A. Taube, M. Gebre-Medhin, and H. Rosling. 2000. Classification of thyroid size by palpation and ultrasonography in field surveys. Lancet 355(9198):106-110.

Rockette, H.E., and V.C. Arena. 1983. Mortality Pattern of Workers in the Niagara Plant. Department of Biostatistics, Graduate School of Public Health, University of Pittsburgh, Pittsburgh, PA. June 1983.

Rothman, K.J., and S. Greenland. 1998. Precision and validity in epidemiologic studies. Pp. 115-134 in Modern Epidemiology, 2nd Ed., K.J. Rothman and S. Greenland, eds. Philadelphia: Lippincott-Raven Publishers.

Rovet, J.F. 1999. Long-term neuropsychological sequelae of early-treated congenital hypothyroidism: Effects in adolescence. Acta Paediatr. 88(432):88-95.

Rovet, J.F. 2002. Congenital hypothyroidism: An analysis of persisting deficits and associated factors. Neuropsychol. Dev. Cogn. Sect. C Child Neuropsychol. 8(3):150-162.

Rovet, J.F. 2003. Long-term follow-up of children born with sporadic congenital hypothyroidism. Ann. Endocrinol. 64(1):58-61.

Schwartz, J. 2001. Gestational Exposure to Perchlorate is Associated With Measures of Decreased Thyroid Function in a Population of California Neonates. M.S. Thesis, University of California, Berkeley.

Soldin, O.P., S.J. Soldin, and J.C. Pezzullo. 2003. Urinary iodine percentile ranges in the United States. Clin. Chim. Acta 328(1-2):185-190.

U.S. Geological Survey. 2004. The National Map Viewer. [Online]. Available at: http://nmviewogc.cr.usgs.gov/viewer.htm [accessed July 8, 2004].

WHO (World Health Organization). 1994. Iodine and Health: Eliminating Iodine Deficiency Disorders Safely Through Salt Iodization: A statement by the World Health Organization. WHO/NUT/94.4. Geneva: World Health Organization.

Zurakowski, D., J. DiCanzio, and J.A. Majzoub. 1999. Pediatric reference intervals for serum thyroxine, triiodothyronine, thyrotropin, and free thyroxine. Clin. Chem. 45(7):1087-1091.

4

Animal Studies

THE quality and relevance of animal studies that have evaluated the effects of perchlorate exposure have been debated. Accordingly, the committee was tasked with evaluation of those published and unpublished studies. In this chapter, the committee reviews the animal studies with particular attention to those critiqued and used to derive the reference dose in the 2002 U.S. Environmental Protection Agency (EPA) draft risk assessment. Because rats have been used as the primary animal model, the committee first compares thyroid function in rats and humans and then discusses animal studies that investigated the effects of perchlorate exposure on serum thyroid hormone concentrations, thyroid histopathology, brain morphometry, neurobehavior, and thyroid tumors. Physiologically based pharmacokinetic models are also reviewed briefly. Because concerns have been raised about the possibility of effects other than those resulting from altered thyroid function, the chapter concludes with a discussion of general toxicologic evaluations with emphasis on immunologic studies. The toxicologic implications of perchlorate's interaction with the sodium (Na^+)/iodide (I^-) symporter (NIS) of other tissues and organs are also discussed.

COMPARISON OF THYROID FUNCTION IN RATS AND HUMANS

The fundamental mechanisms involved in the function and regulation of the pituitary-hypothalamus-thyroid system in rats are qualitatively similar to those in humans (Bianco et al. 2002). However, the dynamics of the two systems differ substantially. The biochemical and physiologic differences between rats and humans related to thyroid function are described in the following paragraphs.

Thyroid hormones in serum are extensively bound to plasma proteins. The proteins that bind thyroxine (T_4) and triiodothyronine (T_3) vary widely among species, and their binding affinities for the thyroid hormones also differ. In humans and other primates, thyroxine-binding globulin (TBG) is the principal protein that binds T_4 (Dohler et al. 1979). It has a very high affinity for T_4: only about 0.03% of the T_4 in serum is in the free unbound form (Hill et al. 1989). Binding sharply reduces clearance of T_4 from serum.

Rats do not have TBG, and most T_4 in rat serum is bound to albumin and transthyretin. The binding affinity of T_4 for TBG is more than a 100 times greater than that of albumin or transthyretin (Hill et al. 1989), and the difference contributes to the higher rate of T_4 clearance in rats. The increased clearance contributes to the need for a higher rate of production of T_4 per unit of body weight in rats to maintain normal concentrations of T_4 (Dohler et al. 1979). The higher production rate is reflected in the histologic appearance of the rat thyroid, which has small thyroid follicles that contain much less colloid than those of primates (McClain 1995). Those features give the rat thyroid a more "functionally active" histologic appearance than that of primates, including humans. The follicular epithelium in rats is cuboidal; that of monkeys appears flattened in comparison. The change in the height of the follicular cells from flattened to cuboidal to columnar represents follicular-cell hypertrophy and is characteristic of the increased functional activity.

There appear to be some differences in the metabolism of T_4 by the liver between rats and humans. Some 50% of T_4 is eliminated via bile in rats, but only 10-15% in humans (Hill et al. 1989). The difference does not reflect a qualitative difference in metabolism, because the major metabolite in bile (glucuronide conjugate) is the same in both species (Hard 1998).

The biochemical and physiologic differences between rats and humans related to the thyroid affect their responses to goitrogens, such as perchlorate. For example, Yu et al. (2002) evaluated inhibition of radioiodide uptake by the thyroid in rats exposed to perchlorate in drinking water at 0, 1.0, 3.0, and 10.0 mg/kg of body weight for 1, 5, and 14 days. After 1 day of perchlorate administration, inhibition of iodide uptake was about 15%, 55%, and 65% at 1.0, 3.0, and 10 mg/kg, respectively. After 5 days, inhibition of iodide uptake was 0, 10%, and 30%. After 14 days, inhibition of iodide uptake was observed only at 10 mg/kg. The data show that the initial inhibition of radioiodide uptake by perchlorate in rats is similar to that in humans. However, rats compensated for the inhibition within 5 days of perchlorate administration, most likely by increasing the expression of NIS in the thyroid. A similar response was not observed in a 14-day human study with perchlorate administration (Greer et al. 2002). The data suggest that compensation occurs more quickly in rats because rats have a smaller reserve capacity of thyroid hormones than humans.

Another example of different responses to perchlorate is related to changes in serum concentrations of thyroid hormones and thyrotropin (thyroid-stimulating hormone, TSH). For example, Siglin et al. (2000) treated male and female rats with ammonium perchlorate at 0.01-10 mg/kg per day. At 14 days, serum T_4 concentrations were significantly decreased at 10 mg/kg per day in both male and female rats. Serum T_3 concentrations were significantly decreased in males at 0.01 mg/kg per day or higher. No significant decreases were observed in serum T_3 concentrations at any dose in female rats. Serum TSH concentrations were significantly increased at 0.20 mg/kg per day or higher in males and at 0.05 mg/kg per day or higher in females.

Studies in adult humans have not found increases in serum TSH or decreases in serum T_4 or T_3 with potassium perchlorate exposure over a similar period. For example, administration of 10 mg of potassium perchlorate per day (0.1 mg/kg of perchlorate per day assuming a 70-kg human) to healthy men for 14 days resulted in no changes in serum thyroid hormone or TSH concentrations during the exposure period (Lawrence et al. 2000). Similarly, Greer et al. (2002) found no decreases in serum thyroid hormones or increases in serum TSH in healthy men and women given perchlorate at up to 0.5 mg/kg per day for 14 days.

There are important differences between rats and humans in pituitary-thyroid function during pregnancy. In humans, serum total T_4 and T_3 concentrations rise progressively—on the average, about 50% during the first trimester of pregnancy—and remain increased during the remainder of pregnancy (Glinoer 1997). That response is due to an increase in serum TBG, which is stimulated by an increase in estrogen production (see Chapter 2). During the first trimester, serum free T_4 and T_3 concentrations also increase slightly because of stimulation of the thyroid gland by chorionic gonadotropin, a hormone produced by the placenta. The primary action of chorionic gonadotropin is to sustain pregnancy, but it also has weak thyroid-stimulating activity. The increases in serum free T_4 and T_3 concentrations decrease TSH secretion slightly in pregnant women. Later in pregnancy, the decrease in production of chorionic gonadotropin results in a return of serum free T_4, T_3, and TSH to concentrations comparable with those in nonpregnant women and in men.

Thyroid hormone concentrations change during pregnancy in rats, but detailed studies are limited to gestation days 17-22. During gestation days 17-22, serum T_4 concentrations in pregnant rats are significantly lower than those in nonpregnant female rats (Fukuda et al. 1980; Calvo et al. 1990;

Versloot et al. 1994). The differences in serum thyroid hormone concentrations between pregnant rats and women are related largely to the lack of TBG and absence of chorionic gonadotropin production in rats.

There are also differences between rats and humans in the timing of development of thyroid function. Thyroid function in rats at birth is relatively immature, equivalent to that of a third-trimester human fetus. The human fetus is protected by maternal thyroid hormone for a longer period of development. In rats, serum total T_4 and T_3 concentrations increase between postnatal days 5 and 15 in association with an increase in a serum thyroid hormone-binding globulin (Obregon et al. 1991). Thereafter, the production of that binding protein decreases, and therefore, serum T_4 and T_3 concentrations fall (Vranckx et al. 1994).

Thus, thyroid function and regulation are qualitatively similar in rats and humans, but important differences in serum thyroid hormone binding and clearance rates and thyroid stimulation by a placental hormone in pregnant women lead to important quantitative differences between the two species. The species differences must be carefully considered in interpreting serum thyroid hormone, TSH, and thyroid histopathology data in studies that use rats to assess human health risk associated with perchlorate exposure.

THYROID HORMONES AND THYROID HISTOPATHOLOGY

The committee reviewed published literature and laboratory study reports on the effects of perchlorate exposure on thyroid hormones, TSH, and thyroid histopathology in animals. The discussion here focuses on the Argus (2001) study because it provides a comprehensive evaluation of those entities in the most sensitive populations—pregnant females, fetuses, and neonates. The findings of the Argus study are generally consistent with those of previous studies. Inconsistencies are most likely due to small sample sizes that were used in some of the other studies or to differences in study design, such as use of animals at different stages of development.

In Argus (2001), female rats (dams) were exposed to perchlorate through gestation and lactation. Ammonium perchlorate was administered in drinking water at concentrations that provided doses of 0, 0.01, 0.1, 1.0, and 30 mg/kg per day. Administration began 2 weeks before mating and extended through postnatal day 22. The offspring were exposed to perchlorate in utero, through their mother's milk, and through any consumption of the perchlorate-containing drinking water provided to their mothers. Serum thyroid hormones and TSH were measured on gestation day 21 in the dams and fetuses, on postnatal days 10 and 22 in the dams, and on postnatal days 5, 10, and 22 in the pups. Histologic thyroid evaluations were conducted on the same schedule as the hormone analyses. Standard toxicologic, reproductive, and developmental end points also were evaluated. Findings regarding neurodevelopmental measures are discussed in the next section.

Changes in thyroid hormones and TSH are illustrated in Figures 4-1 and 4-2. Statistically significant differences between control and perchlorate dose groups as determined by Argus (2001) are indicated in the figures. The changes were consistent with those of previous rat studies and with perchlorate's known mode of action in inhibiting the NIS. There were dose-related increases in serum TSH and dose-related decreases in serum total T_4 and T_3 in the dams, fetuses, and pups. Overall, the dams appeared to be more sensitive to perchlorate administration than the fetuses or pups, and the most dramatic changes in the dams were observed on gestation day 21. Although a downward trend was observed in serum T_3 in the dams, it appeared to be a less sensitive marker than T_4. Serum T_3 was decreased significantly only during gestation and only at the highest dose (30 mg/kg per day), whereas serum T_4 was decreased significantly at all doses during gestation and at the highest dose on postnatal days 10 and 22. An important point is that the serum T_4 levels of control rats on gestation day 21 were substantially lower than those of female rats on postnatal days 10 and 22. Those data are consistent with the literature, as discussed previously, and suggest a low thyroid hormone reserve in pregnant rats.

Another way to evaluate the data is to calculate the percentage change from the control value. Figure 4-3 illustrates the changes from controls in serum TSH, T_4, and T_3 in dams, fetuses, and pups at the different evaluation times. Several points should be noted. First, the greatest changes from control values were observed in TSH in the dams, particularly in the highest-dose group. For example, on gestation day 21, when the greatest changes were observed, TSH was increased by 35% at 0.01 mg/kg per day, 50% at 0.1 mg/kg per day, 64% at 1.0 mg/kg per day, and 146% at 30 mg/kg per day. Second, T_4 and T_3 in the dams differed by only about 10% or less from that in controls at the different evaluation times, with the notable exception of T_4 on gestation day 21, which was decreased by 11% at 0.01 mg/kg per day, 45% at 0.1 mg/kg per day, 48% at 1.0 mg/kg per day, and 54% at 30 mg/kg per day. Third, presentation of the data as percentage changes from the control more clearly illustrates the greater sensitivity of T_3 in the pups than in the dams and the greater sensitivity of perchlorate-associated decreases in serum T_3 than in T_4 in the pups. Fourth, the data from postnatal day 22 indicate that males have a greater reduction than females in total T_4 and T_3 compared with controls.

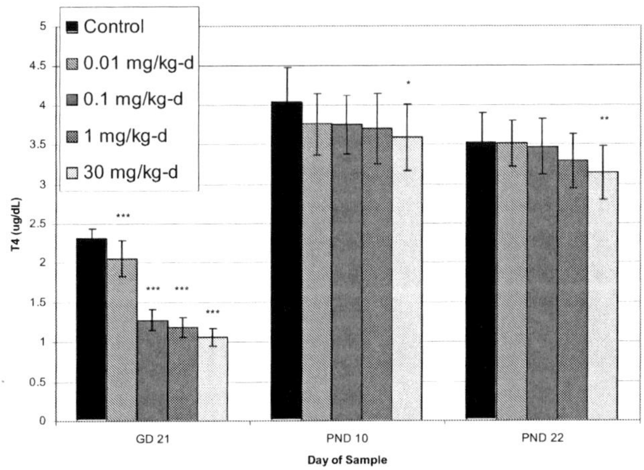
Control
0.01 mg/kg-d
0.1 mg/kg-d
1 mg/kg-d
30 mg/kg-d
T4 (ug/dL)
GD 21
PND 10
PND 22
Day of Sample

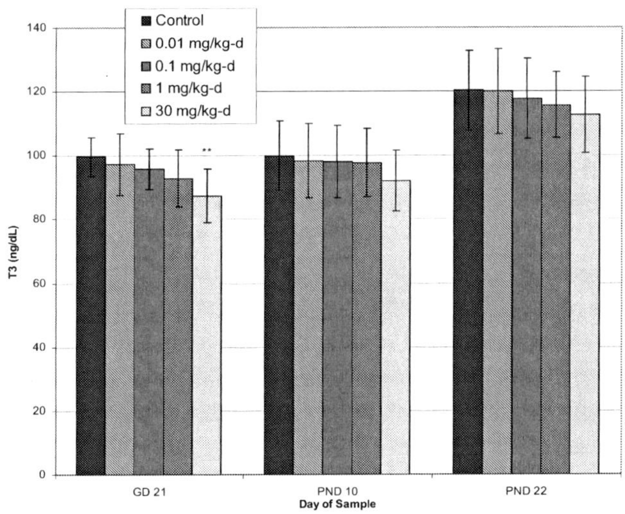
Control
0.01 mg/kg-d
0.1 mg/kg-d
1 mg/kg-d
30 mg/kg-d
T3 (ng/dL)
GD 21
PND 10
PND 22
Day of Sample

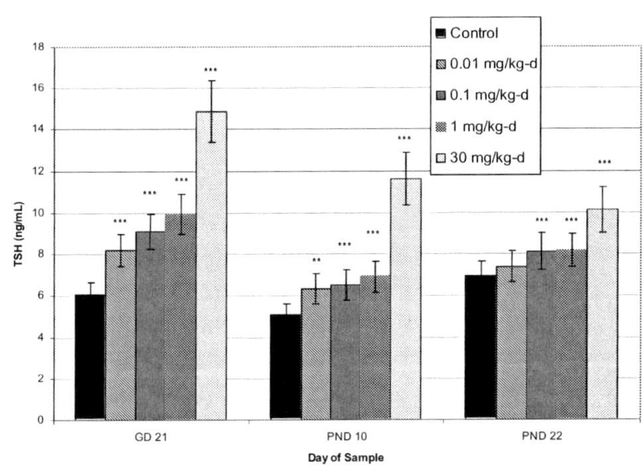
Control
0.01 mg/kg-d
0.1 mg/kg-d
1 mg/kg-d
30 mg/kg-d
TSH (ng/mL)
GD 21
PND 10
PND 22
Day of Sample

Histologic evaluations of the thyroid gland were conducted at the same times as those of thyroid hormones and TSH. Absolute and relative thyroid weights were significantly increased in the dams at 30 mg/kg per day at all evaluation times. A similar trend was noted in absolute thyroid weight in male and female pups over the course of the study. Statistically significant increases in absolute thyroid weights also were observed in all groups of male pups on postnatal day 10 and in females at 1.0 mg/kg per day on postnatal day 22. Histologic examination of the thyroid gland revealed colloid depletion, follicular-cell hypertrophy, and follicular-cell hyperplasia in the dams (see Table 4-1). Those effects were mainly restricted to the highest-dose group (30 mg/kg per day), although colloid depletion and follicular-cell hyperplasia were increased at 1.0 mg/kg per day on postnatal days 10 and 22, respectively. The predominant effect in the fetuses and pups was colloid depletion (see Table 4-2). Colloid depletion was present primarily in the highest-dose group (30 mg/kg per day) but was somewhat increased on several evaluation days in the 1.0-mg/kg group. Follicular-cell hyperplasia was noted occasionally in a few pups, but no clear dose-response trends were noted.

The committee has concerns about the reliability of the thyroid histopathology data, particularly those on the dams. For example, hyperplasia was observed on postnatal day 22 at the lower doses in the absence of hypertrophy, which typically does not occur in rats. The data at the highest dose appeared to be more reliable inasmuch as the expected TSH-mediated morphologic changes in the thyroid were observed: colloid depletion, follicular-cell hypertrophy, and follicular-cell hyperplasia.

The histologic data on the fetuses and pups were more consistent. Colloid depletion of thyroid follicles, although a subjective morphologic end point, was the most consistent histologic finding in rat fetuses and pups on all evaluation days. It was observed consistently in the highest-dose animals and, to a smaller extent, in the 1.0-mg/kg group. The thyroid morphology of the two lower-dose groups of animals (0.01 and 0.1 mg/kg per day) was similar to that of control animals.

FIGURE 4-1 (Top) Serum thyroxine (T_4) concentrations in dams treated with ammonium perchlorate at indicated doses in drinking water. (Middle) Serum triiodothyronine (T_3) concentrations in same animals. (Bottom) Serum thyroid-stimulating hormone (TSH) in same animals. Values presented as mean ± SD of 14-16 animals (data from Argus 2001). Abbreviations: GD, day of gestation; PND, postnatal day; *, significantly different from control, $p \leq 0.05$; **, significantly different from control, $p \leq 0.01$; *** significantly different from control, $p \leq 0.001$.

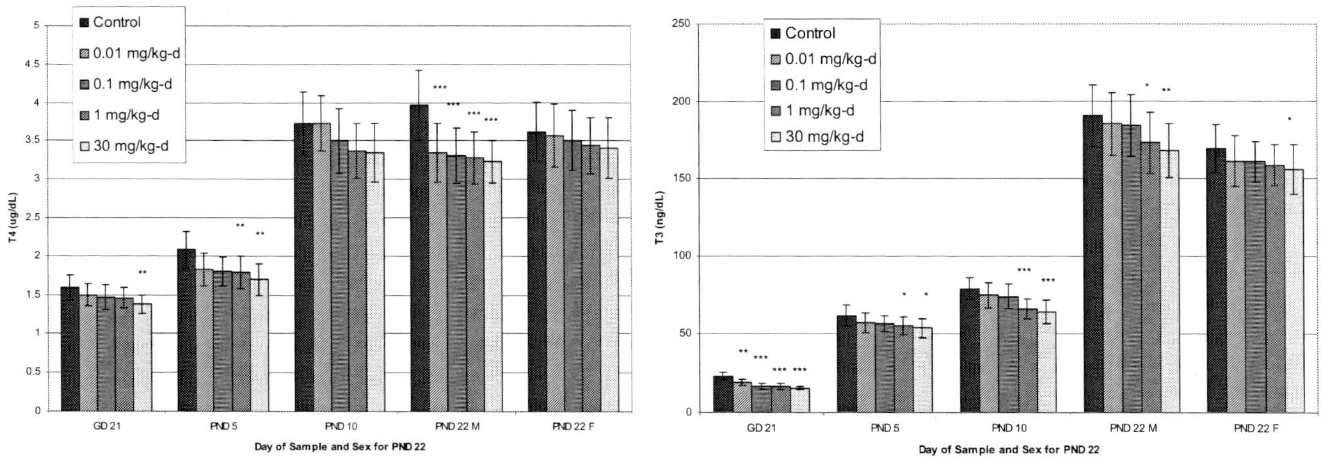

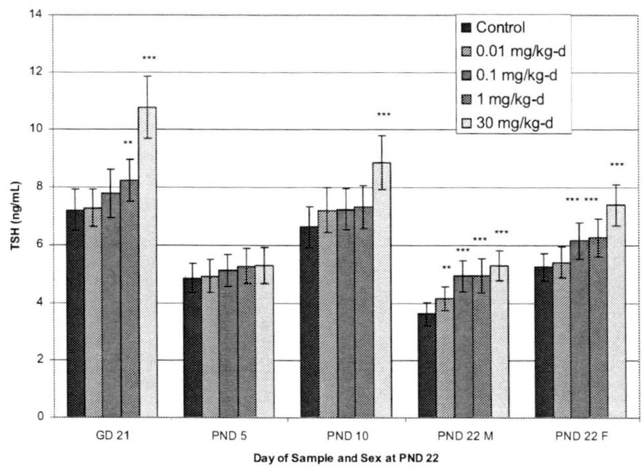

FIGURE 4-2 (Upper Left) Serum thyroxine (T_4) concentrations in fetus and pups of dams treated with ammonium perchlorate at indicated doses in drinking water. (Upper Right) Serum triiodothyronine (T_3) concentrations in same animals. (Lower Left) Serum thyroid-stimulating hormone (TSH) in same animals. Values reported as the mean ± SD of 11-17 animals except for T_3 measures at day 21 of gestation, when two to eight animals were used to derive values (data from Argus 2001). F, female; GD, gestation day; M, male; PND, postnatal day; *, significantly different from control, $p \leq 0.05$; **, significantly different from control, $p \leq 0.01$; ***, significantly different from control, $p \leq 0.001$.

TABLE 4-1 Thyroid Histopathology in Control Dams and Dams Given Four Doses of Perchlorate

	Dose (mg/kg per day)				
Evaluation Day and Effect	0	0.01	0.1	1.0	30.0
Gestation day 21					
Colloid depletion	0/16	0/16	0/16	0/15	16/16
Follicular-cell hypertrophy	0/16	0/16	0/16	0/15	14/16
Follicular-cell hyperplasia	0/16	0/16	0/16	0/15	2/16
Postnatal day 10					
Colloid depletion	0/16	1/16	1/16	5/16	16/16
Follicular-cell hypertrophy	0/16	0/16	1/16	0/16	16/16
Follicular-cell hyperplasia	0/16	0/16	1/16	2/16	9/16
Postnatal day 22					
Colloid depletion	0/16	0/16	1/15	1/16	16/16
Follicular-cell hypertrophy	0/16	0/16	0/15	0/16	14/16
Follicular-cell hyperplasia	3/16	4/16	5/15	10/16	10/16

Source: Data from Argus 2001.

TABLE 4-2 Thyroid Histopathology in Control and Perchlorate-Exposed Fetuses and Pups

	Dose (mg/kg per day)				
Evaluation Day and Effect	0	0.01	0.1	1.0	30.0
Gestation day 21—Fetuses					
Colloid depletion—males	0/16	2/16	0/16	12/16	16/16
Colloid depletion—females	0/16	1/16	1/16	13/16	16/16
Postnatal day 5—Pups					
Colloid depletion—males	0/16	2/16	0/16	4/16	16/16
Colloid depletion—females	0/16	0/16	0/16	6/16	16/16
Postnatal day 10—Pups					
Colloid depletion—males	0/16	1/16	1/16	1/16	16/16
Colloid depletion—females	0/16	0/16	1/16	4/16	15/15
Postnatal day 22—Pups					
Colloid depletion—males	0/16	0/16	0/15	0/16	11/16
Colloid depletion—females	0/16	0/16	0/15	0/15	12/16

Source: Data from Argus 2001.

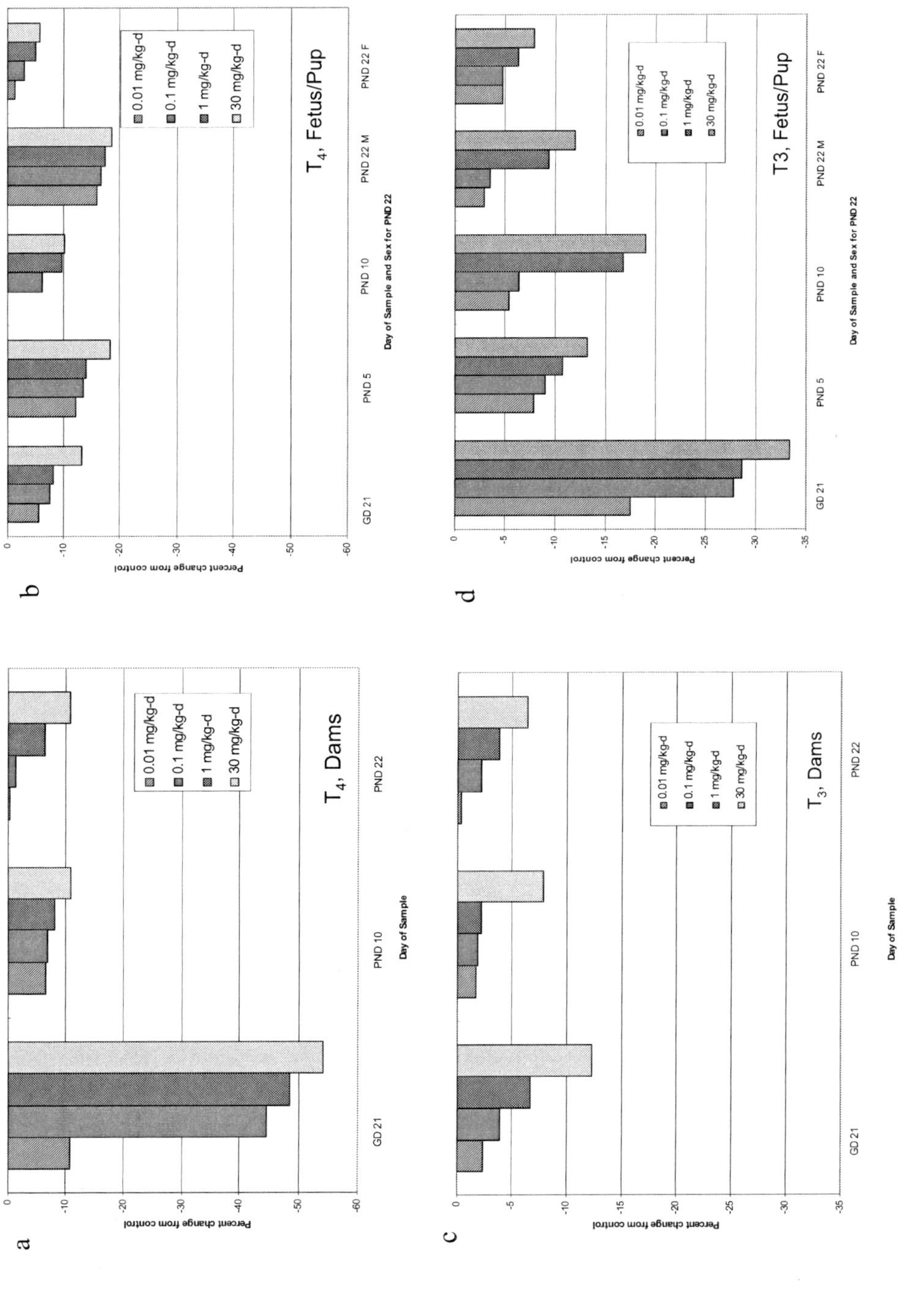
a
T$_4$, Dams
0.01 mg/kg-d
0.1 mg/kg-d
1 mg/kg-d
30 mg/kg-d
Percent change from control
0
-10
-20
-30
-40
-50
-60
GD 21
PND 10
PND 22
Day of Sample
b
T$_4$, Fetus/Pup
0.01 mg/kg-d
0.1 mg/kg-d
1 mg/kg-d
30 mg/kg-d
Percent change from control
GD 21
PND 5
PND 10
PND 22 M
PND 22 F
Day of Sample and Sex for PND 22
c
T$_3$, Dams
0.01 mg/kg-d
0.1 mg/kg-d
1 mg/kg-d
30 mg/kg-d
Percent change from control
0
-5
-10
-15
-20
-25
-30
-35
GD 21
PND 10
PND 22
Day of Sample
d
T3, Fetus/Pup
0.01 mg/kg-d
0.1 mg/kg-d
1 mg/kg-d
30 mg/kg-d
Percent change from control
GD 21
PND 5
PND 10
PND 22 M
PND 22 F
Day of Sample and Sex for PND 22

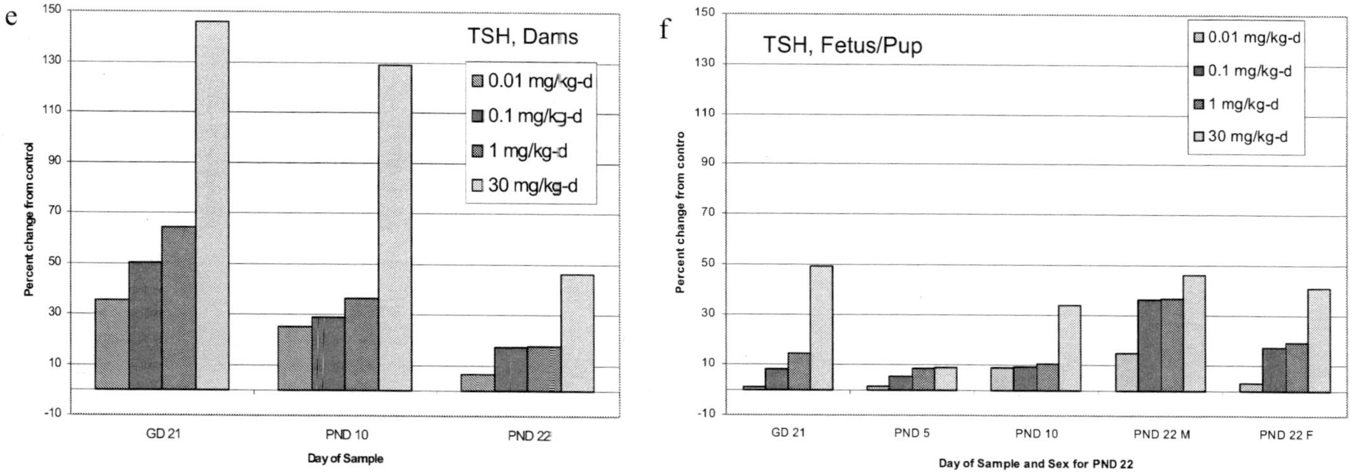

FIGURE 4-3 Changes in serum T_4, T_3, and TSH in dams (a, c, e) and fetuses and pups (b, d, f) presented as percent change from control (data for calculations were those of Argus 2001). Abbreviations: F, female; GD, gestation day; M, male; PND, postnatal day.

Whether changes of the magnitude observed in the Argus (2001) study can cause adverse effects in the animals, particularly the offspring, is the critical issue. EPA (2002a) provided no discussion of the probable implications of the changes observed. The pregnant dams had decreases greater than 10% in serum T_4 at 0.1 mg/kg per day and above. However, it is important to emphasize that the variations in serum T_4 in the control animals at different stages were larger than the changes induced by perchlorate. Serum TSH was substantially increased in the pregnant dams at all doses. Although statistically significant changes in the thyroid hormones and TSH were noted in the fetuses and pups, the changes tended to be more modest than those in the dams.

BRAIN MORPHOMETRY

Because thyroid hormones play a critical role in brain development, two Argus (1998, 2001) studies included measures of brain morphometry. In the 1998 study, female Sprague-Dawley rats were exposed to ammonium perchlorate at 0, 0.1, 1.0, 3.0 or 10.0 mg/kg per day in the drinking water beginning on gestation day 0 and continuing until postnatal day 10. One male and one female pup from each of six control and six high-dose litters were sacrificed for brain morphometry on postnatal day 10-12 and postnatal day 82-85. The brains taken on day 10-12 were immersion-fixed in 10% neutral buffered formalin, embedded in paraffin, and sectioned coronally. Linear measurements were taken from three sections. A section through the optic chiasm was used to measure the thickness of the frontal and parietal cortex, the diagonal width of the striatum, and the thickness of the corpus callosum. The thickness of the hippocampal gyrus was measured in a section taken just posterior to the mammillary body. A section taken just anterior to the midpoint of the cerebellum was used to measure the height of the cerebellum at the level of the deep cerebellar nuclei, including lobes 1-6 and extending from the roof of the fourth ventricle to the dorsal surface.

Initially, only the brains from the 0- and 10.0-mg/kg groups were sectioned and measured. The only statistically significant difference reported by Argus for the day 10-12 brains was an increase in the thickness of the corpus callosum in the female pups. An increase of similar magnitude was observed in the male pups, but the data on the male pups were more variable, and the difference was not statistically significant. After reviewing the data, EPA requested that sections from the 3.0-mg/kg group be assessed. Apparently, brains from the 0.1- and 1.0-mg/kg groups were never assessed. EPA then performed statistical analyses to compare the 0-, 3.0-, and 10.0-mg/kg groups. EPA reported a significant increase in the thickness of the corpus callosum in the 10.0-mg/kg group, significant decreases in the thickness of the hippocampal gyrus and striatum in the 3.0-mg/kg group, and significant increases in the thickness of the anterior and posterior cerebellum in the 3.0-mg/kg group. No alterations in the thickness of the corpus callosum were observed at 3.0 mg/kg per day. It does not appear that those effects were sex-specific, although the EPA report is not clear on this point (EPA 2002a; p. 5-37).

The brains from control and 10-mg/kg group on postnatal day 82-85 were perfusion-fixed in 10% neutral buffered formalin and later sectioned and measured according to the same method used with the postnatal day 10-12 brains. Statistically significant increases in the thickness of the corpus callosum and frontal cortex and in the cross-sectional width of the striatum were observed in the exposed males, but not in the females. Apparently, brains from the 3.0-mg/kg group were not assessed on postnatal day 82-85. Significant changes in brain morphometry at 10-12 and 82-85 days of age are summarized in Table 4-3.

The neuromorphologic findings from Argus (1998) are difficult to interpret because of the absence of a dose-effect relationship and the inconsistency of effects across age and sex. A possible exception is the increase in the thickness of the corpus callosum, which was present in the highest-dose male rats (10 mg/kg) at both ages, although the difference did not reach statistical significance on postnatal day 10-12.

Because of the difficulties in interpreting the data from the 1998 study, Argus (2001) conducted additional neuromorphometric analyses. Perchlorate exposure of female Sprague Dawley rats started 2 weeks before cohabitation with male rats and continued until they were sacrificed. Doses of 0, 0.01, 0.1, 1.0, and 30.0 mg/kg per day were used, and pups were sacrificed for brain morphometry on postnatal days 10 and 22. The brains were immersion-fixed, embedded, and sectioned with the same methods used in the 1998 study. Argus (2001) reported a number of statistically significant differences between treatment groups, including increases in the thickness of the corpus callosum of male rats on postnatal days 10 and 22, finding that are consistent with those in the earlier study. However, there was not a clear dose-response relationship. In the 1998 study, an increase in the thickness of the corpus callosum was seen at 10 mg/kg per day but not at 3 mg/kg. In contrast, in the 2001 study, increases in corpus callosum thickness were seen at 0.1 and 1.0 mg/kg per day but not at 30 mg/kg.

There is some confusion as to how the data were statistically analyzed by Argus. EPA states that the differences were determined by performing multiple t tests (EPA 2002a; p. 5-65). However, the Argus report (Argus

TABLE 4-3 Summary of Morphometric Findings in Rat Pups Exposed to Perchlorate[a]

Neuroanatomic	Argus 1998		Argus 2001		EPA 2003
Region	Day 10-12	Day 82-85	Day 10	Day 22	Day 22
Frontal cortex	No change	Increase at high dose only in males	Increase at low or intermediate doses but not at high dose	Increase at high dose only in males	Not measured
Parietal cortex	No change	No change	Increase at low or intermediate doses but not at high dose	No change	Not measured
Striatum	Decrease at low or intermediate doses but not at high dose	Increase at high dose only in males	Increase at low or intermediate doses but not at high dose	Decrease at low or intermediate doses but not at high dose	All dose groups increased
Corpus callosum	Increase at high dose only	Increase at high dose only in males	Increase at low or intermediate doses but not at high dose in males; Decrease at low or intermediate doses but not at high dose in females	Increase at low or intermediate doses but not at high dose	Increase at low or intermediate doses but not at high dose
Hippocampus gyrus	Decrease at low or intermediate doses but not at high dose	No change	No change	No change	Not measured
Dentate gyrus	Not measured	Not measured	No change	No change	Not measured
CA1 portion	Not measured	Not measured	Decrease at low or intermediate doses but not at high dose in females; Increase at low or intermediate doses but not at high dose in males	No change	Not measured

CA3 portion	Not measured	Not measured	No change	Decrease at low or intermediate doses but not at high dose	Not measured
External germinal layer of cerebellum	Not measured	Not measured	Decrease at low or intermediate doses but not at high dose	No change	Not measured
Anterior/posterior cerebellum	Increase at low or intermediate doses but not at high dose	No change	No change	Increase at low or intermediate doses but not at high dose	Not measured

[a]Doses assessed at different evaluation points in

Argus (1998) – day 10-12 (0, 3, and 10 mg/kg per day); day 82-85 (0 and 10 mg/kg per day)

Argus (2001) – day 10 (all dose groups); day 22 (all dose groups).

2001, pp. 43-44 and p. 820) states that one-way ANOVAs were performed and were followed by Dunnett's comparisons of each treatment group with the control group when the overall ANOVA was significant. In either case, a large number of statistical comparisons were run, so Type I error is a concern.

EPA later reanalyzed the same data using two statistical approaches. In the first, it used a multivariate approach known as profile analysis. Briefly, profile analyses make comparisons between groups by creating a vector composed of all the measurements taken from an individual animal. The primary test—or test of parallelism of vectors—establishes whether the pattern of results across brain regions differs between treatment groups. Data from the right and left hemispheres were averaged for those analyses. Male and female pups were included in the same analysis with sex nested within a litter, but separate profile analyses were run on the data from postnatal days 10 and 22. In the second approach, univariate ANOVAs and Dunnett's comparisons were used to determine which brain regions and dose groups differed from controls. That approach was similar to the one used by Argus, except that data on the two hemispheres were averaged rather than analyzed separately and the litter was considered the unit of variance with sex as a nested variable within a litter.

The profile analysis indicated a significantly different pattern of results across treatment groups. The finding was due primarily to changes in the perchlorate-exposed groups compared with the control group in thickness of the striatum (decrease), posterior corpus callosum (increase), and cerebellum (increase). For several of the observed effects, the dose-effect relationship was an inverted U-shaped function with the low or intermediate dose groups showing the largest deviations from controls (see EPA 2002a, Figure 5-15, p. 5-69). The results of the univariate analysis of the same data are summarized in Table 4-3 (see column labeled Argus 2001, Day 22). That analysis revealed a pattern of effects similar to that revealed by the profile analysis. However, several additional significant findings were observed, including an inverted U-shaped dose-effect function in region CA3 of the hippocampus and an increase in the thickness of the frontal cortex that was present mainly in the male pups.

A number of outside experts have reviewed the Argus studies and raised substantial questions and concerns regarding the study design, the methodology, and the biologic plausibility of the findings. Methodologic problems that have been discussed include method of fixation, time since fixation, variation in the plane of section, and the conduct of some measurements in a nonblinded fashion. Various reviewers have also pointed out the relative insensitivity of linear measurements of thickness relative to measurements of the area or volume of a structure and have raised questions about the appropriateness of using coronal sections to measure the corpus callosum. Finally, reviewers have raised concerns about the lack of a clear dose-response relationship and questioned whether reductions in thyroid hormones that were not large enough to reduce the growth rate or overall brain weight of the pups would be large enough to produce gross alterations in neurodevelopment.

The Argus studies also have been criticized for using immersion fixation rather than perfusion to fix pup brains for sectioning (TERA 2001; Elberger 2003). There does not appear to be a clear consensus among experts on that point. Harry (2001) notes that immersion fixation is appropriate for brains of young pups because the decreased blood volume in the pup can result in less than optimal perfusion. However, Juraska (J. Juraska, University of Illinois, Champaign, personal commun., 2004) states that although that is true for pups under 5 days old, perfusion would be the most appropriate method of fixation for brains of 10- or 22-day-old pups. TERA (2001) notes that with immersion fixation distortion of the tissue can occur as the brain fixes from the outside in. Because it takes longer, immersion fixation can also lead to degeneration of cellular components. That makes the brain more susceptible to shrinkage and swelling. TERA (2001) notes that brains fixed with any method shrink during the first several weeks in fixative and swell later. Therefore, it is critical that the time between fixation and sectioning be held constant across treatment groups in a study, particularly if immersion fixation is used. All tissues from the Argus studies were embedded in paraffin at the same time, but tissues from some groups were held longer than others before being sectioned and stained. Because all tissues from each Argus study were embedded in paraffin at the same time, differences in how long the tissues were held before sectioning should not be an issue.

In addition to questions about fixation of the brains, reviewers have noted what appears to be considerable variability in the plane of section—both between animals and between hemispheres in the same animal—in both Argus studies (Harry 2001; Wahlsten 2002). Variability, if equally distributed across dose groups, will increase Type II error, making it more difficult to detect treatment-related effects, but it should not increase the likelihood of spurious effects (Type I error). No evidence has been presented to suggest that the anterior-posterior coordinates of the sections varied systematically by dose group in the 1998 study. However, both Harry (2001) and Wahlsten (2002) have reviewed brain sections from the 2001 study in detail and have concluded that differences in the plane of section varied systematically across groups in that study. Wahlsten (2002) has noted (1) that brains with a thicker corpus callosum appear to have been sectioned more posteriorly near the splenium of the corpus callosum, which is its thickest part, and (2) that there were more of these posterior "near splenium" sections in the three low-dose groups. Harry (2001) reviewed the sections used to measure the thickness of the corpus callosum with reference to a stereotaxic atlas of the rat brain and noted that although only 10% of the sections from the control group were taken above anterior-posterior coordinate 3.5, 23% and 56% of the sections from the 0.1- and 1.0-mg/kg groups were cut at anterior-posterior coordinates above 3.5. Those were the groups that showed increases in corpus callosum thickness relative to the control group. Harry (2001) has noted that the differences in the plane of section could also influence measurements of the striatum and frontal cortex.

A further criticism of Argus (2001) is that morphometric measurements of the sections from the three low-dose groups—which were sectioned and measured later than the control and high-dose groups—were not conducted in a blinded fashion. The committee shares that concern. It is important to distinguish between comparative-morphology studies and morphometric studies. Comparative morphology yields *qualitative* assessments of the appearance of tissues; specifically, tissues of control animals are compared with those of exposed animals to determine whether specific pathologic changes are present in the exposed group. Morphometric studies yield *quantitative* measures of linear thickness of tissues. In Argus (2001), linear thickness of various brain regions was measured. Although it may sometimes be appropriate to collect data in a nonblinded fashion in a comparative-morphology study, there is no defensible reason for collecting data in a nonblinded fashion in a morphometric study.

To address the concern about differences in the plane of section across animals, EPA contracted with Consultants in Veterinary Pathology, Inc. in 2003 to resection and remeasure brains from the postnatal day 22 pups from Argus (2001). The new sections were matched to anatomic atlas plates to ensure consistency in the plane of section among animals. Only the striatum and anterior and posterior corpus callosum were measured in the follow-up study. Because there was no controversy over plane of section for the cerebellum, measurements of the cerebellum from the original Argus (2001) dataset were included in the statistical analysis. The data were subjected to a profile analysis similar to that performed on the original 2001 dataset. The analysis was limited to a subset of the original 16 animals per group for which tissue was available for the resectioning in 2003. Measurements from 15 control animals, nine 0.01-mg/kg animals, four 0.1-mg/kg animals, 10 1.0-mg/kg animals, and 11 30.0-mg/kg animals were available. The 0.1-

mg/kg group was excluded from the statistical analyses because fewer than the minimum of six animals required by the EPA Developmental Neurotoxicity Testing Guidelines were available (EPA 1998a).

As in the original dataset, the profile analysis was significant and indicated that the pattern differed among treatment groups. All three of the brain regions—cerebellum, striatum, and corpus callosum—were thicker in the treated groups than in the control group (Geller 2003, Figure 3). When the older data from the cerebellum were excluded from the analysis, the treatment effect remained significant.

That corpus callosum thickness was increased at multiple doses and at several stages of development (postnatal days 10-12, 22, and 82-85) in two studies is suggestive of a relationship between perchlorate exposure and altered neurodevelopment. EPA's analysis of additional tissue from Argus (2001) animals helps to dispel some of the concerns that have been raised about systematic differences in the plane of section among treatment groups. However, it also has contributed to the inconsistencies in the dataset because the thickness of the striatum was decreased in the 2001 brain sections and increased in the 2003 brain sections from the same pups.

An important issue that several critiques of the Argus studies (e.g., TERA 2001) have raised is that linear measurement of the thickness of brain areas is not the most sensitive method of detecting alterations in neural structure. Measurement of volume or area would be more sensitive and more accurate, particularly for structures, such as the corpus callosum and hippocampus, that change shape across serial coronal sections. For those structures, linear thickness will depend on the location of the section, and small changes in the plane of section could have a large effect on the results. Several experts agree that a more appropriate method of assessing changes in the size of the corpus callosum would be to measure the area of the structure in a midsagittal section through the brain (TERA 2001; Wahlsten 2002; Elberger 2003). However, it is important to note that all measurements of size, including measurements of volume or area, are no more than a surrogate for changes in the cellular structure of a brain region, and it is ultimately the underlying changes that are important to understand.

Laboratory rats have been used extensively as a model to study the effects of thyroid hormone deficiency or excess on brain development, and there is a large literature on the neuroanatomic effects of neonatal hypothyroidism in rats that should be used to inform the design of future studies of perchlorate. In particular, the laminar organization of layered structures—including the cerebellum, hippocampus, and cortex—is disrupted by neonatal hypothyroidism (see, for example, Alvarez-Dolado et al. 1999). Neonatal hypothyroidism appears to alter the development of those structures by reducing the expression of reelin, a protein that is involved in the radial migration of neurons (Alvarez-Dolado et al. 1999). That results in aberrant migration and blurring of otherwise clearly defined layers in those structures. Another classic sign of neonatal hypothyroidism is stunting of the dendritic trees of pyramidal cells in the hippocampus (Rami et al. 1986) and of Purkinje cells in the cerebellum (Legrand 1986). The changes can be dramatic and are easily visualized with Golgi stain (Rami et al. 1986). Finally, disappearance of the external granular layer of the cerebellum is delayed in neonatal hypothyroidism (Lauder 1977). Analysis of some of those classic signs of neonatal hypothyroidism in perchlorate-exposed animals will be important in establishing whether a pattern of neuromorphologic effects consistent with neonatal hypothyroidism exists.

Alternative approaches could also be useful. A number of thyroid hormone-responsive genes have been identified in the brain (Zoeller et al. 2002). They are differentially expressed in different brain regions at different stages of development. For example, neurotrophin and brain-derived neurotrophic factor gene expression are altered by thyroid status (Koibuchi et al. 2001). Analyses of the effects of perchlorate exposure on the expression of those genes could provide additional information about whether the developing brain responds to perchlorate exposure in a way that is consistent with the response to other manipulations that result in low thyroid hormone concentrations.

In its draft risk assessment (EPA 2002a), EPA summarizes all the morphologic data from the Argus studies but focuses on the effect of perchlorate on the corpus callosum. Several reviewers stated that perchlorate is believed to exert its effects by reducing the production of thyroid hormones, and the corpus callosum is not known to be a target of thyroid hormone. However, although the corpus callosum is not widely recognized as sensitive to thyroid hormones, several studies indicate that maturation of the corpus callosum is under the influence of thyroid hormones. The studies suggest that hypothyroidism may arrest callosal axons at an immature stage of development (Gravel and Hawkes 1990; Berbel et al. 1993). In other words, callosal axons that normally withdraw during development appear to be retained after neonatal hypothyroidism, which could explain an increased callosal thickness. More recent research suggests that the effects may be caused by the inappropriate expression of proteins TAG-1 and L1 (Alvarez-Dolado et al. 2000, 2001). Those proteins are members of the immunoglobulin super family of cell adhesion molecules and are thought to play an important role in axon growth and guidance during development. Alvarez-Dolado et al. (2000, 2001) found that both TAG-1 and L1 are overexpressed in major fiber tracts, including the anterior commissure and the corpus callosum, of hypothyroid rats during development. Therefore, although not widely recognized as a classic marker of neonatal hypothyroidism, increased thickness of the corpus callosum appears to be a biologically plausible effect.

Another important issue related to the biologic plausibility of effects on brain structure is whether perchlorate exposure reduces maternal or pup thyroid hormone concentrations enough to alter neurodevelopment. Most studies of neonatal hypothyroidism in animals have used excessive doses of potent goitrogens that reduce serum T_4 concentrations to near the detection limit of the assay. The potential for modest reductions in serum T_4 to alter gross morphologic end points, such as the thickness of the corpus callosum or frontal cortex, is unknown and may need to be investigated with the use of positive control groups exposed to moderate doses of model goitrogens.

In summary, two studies (Argus 1998; Argus 2001) were conducted in Sprague-Dawley rats to investigate the effect of maternal perchlorate exposure on offspring brain development. Dams were exposed to various doses of perchlorate in drinking water throughout gestation and lactation. Pups were sacrificed at several postnatal ages, and their brains were fixed and sectioned for histologic evaluation. The thickness of various brain regions was measured. Statistical analyses revealed a number of significant effects, most notably an increase in the thickness of the posterior corpus callosum. However, questions and concerns about the studies have been raised, including (1) apparent systematic differences in the plane of section among treatment groups, (2) lack of a clear and consistent dose-response relationship, (3) doubts about the biologic plausibility of the changes that were observed, and (4) concerns that the measures that were used were relatively insensitive and would be unlikely to pick up subtle differences in neurodevelopment. On the basis of its review of the data, the committee concludes that the evidence in Argus (1998) and (2001) is inadequate to determine whether or not a causal relationship exists between maternal perchlorate exposure and pup neurodevelopmental abnormalities.

NEUROBEHAVIORAL STUDIES

The 1998 Argus study included a battery of behavioral tests that included tests of passive avoidance, water-maze learning, motor activity, and auditory startle. The behavioral tests were not repeated in the 2001 Argus study. Auditory-startle habituation amplitude was evaluated on postnatal days 23 and 61. It was an appropriate screening test for the effects of perchlorate because previous studies had shown that auditory-startle amplitudes are altered after exposure to propylthiouracil (PTU), a model goitrogen (Goldey et al. 1995b), or polychlorinated biphenyls (PCBs), a thyroxine-suppressing environmental contaminant (Goldey et al. 1995a). Neonatal exposure to PTU and PCBs resulted in suppressed startle amplitudes at 24 days of age and increased amplitudes in adulthood. In the 1998 Argus study, no significant effects of perchlorate on startle amplitude were noted at either age.

Two learning tests, passive avoidance learning and a water-filled M maze, also were administered. Learning deficits have been reported in children and in animals that had hypothyroidism (Brosvic et al. 2002; Rovet 2002), so inclusion of the tests was appropriate. Passive-avoidance testing conducted on days 23-25 consisted of placing the rats in the bright side of a two-compartment chamber. The rats were allowed to explore until they entered the dark compartment, where a 1-mA electric shock was delivered. The rats were then removed from the dark compartment and placed back in the bright compartment. Testing continued until the rats remained in the bright side of the chamber for 60 seconds in two consecutive trials. No significant effects of perchlorate were noted. Water-maze testing conducted on days 59-63 consisted of a series of trials in which the rats were placed in a standard start position and had to swim to one of two goals. Rats were trained to a criterion of five consecutive errorless swims. Again, no significant effects of perchlorate were noted.

As discussed above, neonatal hypothyroidism results in abnormal development of various brain regions. The hippocampus and cerebellum are two of the most severely affected regions, so such learning tests as the radial-arm maze, which is sensitive to hippocampal damage (Becker et al. 1980), and eye-blink conditioning, which is sensitive to cerebellar damage (Steinmetz 1996), could potentially pick up subtle effects of altered thyroid function on cognition that might be missed by the tests that were administered.

The tests that were used would not be the most appropriate ones for identifying functional effects of alterations in the structure of the corpus callosum, the most striking (although controversial) neuromorphologic findings reported in Argus (1998, 2001). In humans, a thicker corpus callosum has been linked to deficits in executive function (Bookstein et al. 2002), so tests of functions, such as working memory, attention, cognitive flexibility, and inhibitory control, might be useful. Examples would be a delayed-response task, a serial-reversal learning task, and an operant schedule that reinforced animals for low rates of responding. Those tests and the ones mentioned above require considerable time and effort and would be difficult to incorporate into large-scale projects like the Argus studies, which were designed to assess multiple heath outcomes. Furthermore, they are not required in EPA's Developmental Neurotoxicity Testing Guidelines (EPA 1998a). However, without additional cognitive assessments that use tests that focus on the specific cognitive domains most likely to be affected by subtle to moderate reductions in thyroid hormone concentrations, it is impossible to rule out an effect of perchlorate on cognition.

Another useful approach would be a transgenic mouse model with a phenotype characterized by enlarged corpus callosum axons (Seeger et al. 2003). Characterization of the behavioral profile in such animals would provide valuable information that would help to correlate behavioral changes with morphologic effects of perchlorate exposure.

Locomotor-activity testing was conducted at the ages of 14, 18, 22, and 61 days in the 1998 Argus study, and at the ages of 14, 18, and 22 days by Bekkedal et al. (2000). The design of the Bekkedal et al. (2000) study was similar to that of the Argus study. Female Sprague-Dawley rats were treated with ammonium perchlorate at 0, 0.1, 1.0, 3.0, or 10.0 mg/kg per day in the drinking water beginning 2 weeks before gestation and continuing until postnatal day 10. The rats in both studies were tested in an open-field activity monitor transected by infrared photobeams. The screening tool was appropriate because treatment with PTU or PCBs has been shown to suppress locomotor activity during preweaning development and increase locomotor activity at later ages (Goldey et al. 1995a, 1995b). However, locomotor activity can be variable because it is influenced by many parameters. Therefore, it is not a very sensitive measure of exposure-related effects. No significant effects were noted in the Bekkedal et al. (2000) study. The only significant effect noted in the Argus study was an increase in activity on postnatal day 14 in male pups in the highest-exposure perchlorate group (10.0 mg/kg). The effect was manifested as a failure to habituate over the 1.5-hour test session. Argus's statistical analyses indicated a significant effect when male and female pups were evaluated separately, but a reanalysis in which male and female data were combined found no significant effects.

In November 2001, David Dunson, of the National Institute of Environmental Health Sciences (NIEHS), was asked to provide an analysis of Argus (1998) and of Bekkedal et al. (2000) because the original contractor was unable to find statistically significant effects despite large increases in several outcome variables. EPA neurotoxicologists believed that the increases were of concern from a biologic perspective. Because of the high correlation between different types of motor-activity measurements, Dunson focused on the number of ambulatory movements, which was viewed as most reflective of overall motor activity. He used a Bayesian hierarchic model to assess the dose-response trend in motor activity. In a Bayesian statistical approach, expert judgment and available data are modeled in the prior-probability distribution. When data and expertise are scarce, the prior-probability distribution is diffuse (spread over a wide range of plausible values, each with low probability). When the prior distribution is developed, Bayes's theorem is used to update it with information from the current data to arrive at the posterior-probability distribution. In many cases, implementation of Bayes's theorem involves complex integrals. In those complex modeling situations, Markov Chain Monte Carlo methods can be used to make calculation of Bayes's theorem simple by using sequential sampling algorithms. Dunson (2001) included dose, sex, age, habituation time, and a habituation-time-by-dose-interaction term in the model. The correlation due to the repeated-measure structure of the data was accounted for by using an animal-specific intercept. Plausible but diffuse prior distributions were used for all parameters. The prior distribution of most interest, that for the dose-response relationship, was centered on a value corresponding to no effect of perchlorate exposure. The committee observed that the Bayesian model was carefully designed and that the analysis was properly conducted according to standard procedures.

Dunson's reanalysis of the Bekkedal data yielded a posterior probability of 98% that motor activity increased with perchlorate exposure. The probability increased from 79% in the first habituation interval to over 99% in the final interval. The dose estimated to increase the mean number of ambulatory movements at the final habituation time by 10% (effective dose 10%, or ED10) was 1.62 mg/kg per day with a 95% credible interval[1] of 0.90-7.87 mg/kg per day. Dunson's reanalysis of the Argus data yielded a high posterior probability that the motor activity increased with perchlorate exposure. The probability increased from 58% in the first habituation interval to 94% in the final interval. The ED10 was 4.60 mg/kg per day with a 95% credible interval of 2.18 mg/kg per day to infinity. To combine the two studies, the model was modified to include distinct baseline parameters for the error variances, intercept, age effects, and habituation-time effects, but the slopes for the dose-response relationship were assumed to be common across studies. In the combined analysis, there was a posterior probability of 99% that motor activity increased with perchlorate exposure in the final habituation interval. The ED10 was 3.33 mg/kg per day with a 95% credible interval of 1.91-12.78 mg/kg per day.

---

[1]A Bayesian credible interval is an interval such that the probability that the parameter lies in the interval is at least the given percentage.

Two concerns arise with Dunson's analysis. The first is derived from the fact that the point estimate of the ED10 from the Bekkedal et al. (2000) study was 1.62 mg/kg per day and does not fall into the posterior credible interval of the combined analysis. Although Dunson was using a "standard" method for the day, it is likely that he underestimated the uncertainty in the ED10 by his choice of distributions for the combined analysis. Underestimating uncertainty is an example of problems with the "standard" that statisticians have come to recognize (Berry 2000). The second concern is derived from using the lower bound of the credible interval to come up with conservative estimates. Indeed, whenever the sample is small, intervals will be wide, and lower bounds will be small. Although Dunson presents credible intervals, presenting only conservative estimates based on lower bounds is misleading. Dunson claims that an ED10 of 1-2 mg/kg per day represents a conservative estimate, but it is over-conservative. On the basis of his own analysis, the posterior probability that the ED10 is less than or equal to 2 mg/kg per day would be very small; therefore, the posterior probability that the ED10 would be greater than 2 mg/kg per day would be large.

As with the other measures that were used to assess behavioral function, general motor activity is not necessarily the most relevant or most sensitive aspect of motor function to assess if neonatal hypothyroidism is the suspected mechanism of action. As discussed above, the cerebellum—a brain region that is critically important for controlling balance and coordination—is severely affected by neonatal hypothyroidism. Such tests as the rotating rod and rope climb, which were devised to assess the functional consequences of cerebellar damage (Altman and Bayer 1997), are relatively easy to conduct and may be more sensitive to effects of perchlorate exposure. Those tests have been used to assess motor function in rats exposed to alcohol, which is known to target the cerebellum (Klintsova et al. 2000), or PCBs, which are known to decrease serum T_4 concentrations (Roegge et al. 2004).

Another functional end point that may be important to assess in perchlorate-exposed pups is hearing. Thyroid hormones are known to play an important role in the development of the cochlea, and neonatal hypothyroidism has been linked to hearing loss in animals (Knipper et al. 2000) and humans (Wasniewska et al. 2002). PCBs also cause hearing loss (Goldey et al. 1995b), which can be at least partially ameliorated by concurrent treatment with T_4 (Goldey and Crofton 1998). The deficit appears to result from damage to the outer hair cells of the cochlea (Crofton et al. 2000). Recent studies suggest that a moderate decrease in T_4 (about 50%) is sufficient to cause significant low-frequency hearing loss. Thus, the moderate decreases in T_4 observed by Argus (1998, 2001) may be large enough to cause subtle deficits in auditory function.

Hearing can be assessed in rodents with a number of methods, including reflex-modification audiometry (e.g., Goldey et al. 1995a,b) and the measurement of distortion-product otoacoustic emissions (Lasky et al. 2002). The latter requires fairly sophisticated equipment but has several advantages. First, a detailed assessment of hearing across a broad range of frequencies can be obtained from an anesthetized animal in a single 1-hour test session. Second, the method directly assesses the functional integrity of the cochlea, a primary site of damage in hypothyroid-induced hearing loss (Ng et al. 2004). Third, the method can be applied in animals and human infants, facilitating extrapolation of findings from animals to sensitive human populations. A recent study with the method found hearing loss across a range of low and middle frequencies in PCB-exposed animals (Widholm et al. 2003).

Future rodent studies should focus on the specific functional end points described above that are likely to be adversely affected by moderate reductions in thyroid hormone. Those include balance and coordination and motor learning (Altman and Bayer 1997), spatial working memory (Olton et al. 1979), and auditory function (Lasky et al. 2002). The behavioral and sensory tests should be accompanied by studies of brain morphology that go beyond simply measuring the thickness, area, or volume of particular brain areas and look for neuromorphologic changes in the cerebellum, hippocampus, and cochlea that are consistent with neonatal hypothyroidism. The studies should make use of specialized staining techniques—such as the Golgi, Timms, and Nissl stains—which can be used to visualize changes in cerebellar Purkinje and hippocampal pyramidal cell morphology (Gould et al. 1990; Golgi stain), hippocampal mossy fiber pathways associated with spatial learning and memory (Schwegler et al. 1990; Timms stain), and general neuronal organization and neuronal cell counts in various brain regions (Ribak, 1986; Nissl stain). Outer hair cells of the cochlea should also be assessed because they are damaged when insufficient thyroid hormone is present during development (Crofton et al. 2000). For those approaches to be successful, it is imperative that the brains be perfusion-fixed before sectioning and staining and that the assessments be conducted in a blinded fashion.

Relationships between corpus callosum alterations and behavioral sequelae have been described (Magara et al. 2000), so the biologic significance of changes in the corpus callosum should not be dismissed. As described above, a transgenic mouse model with a phenotype characterized by enlarged corpus callosum axons (Seeger et al. 2003) may be of use in determining behavioral deficits that may be associated with enlargement of the corpus callosum. The p21H-Ras gene is associated with that transgenic phenotype; looking at changes in expression of this gene after treatment with perchlorate may provide additional means of correlating the corpus callosum changes with behavioral changes and may help validate that correlation.

It would also be useful to measure the extent to which perchlorate exposure during brain development alters the expression of thyroid hormone-responsive genes and gene products in the brain. That approach has been useful in understanding the actions of other agents that affect the thyroid system (Zoeller et al. 2002; Gauger et al. 2004).

In summary, a number of behavioral measures were assessed in the 1998 Argus study. Motor activity, auditory startle, and learning and memory were appropriate functions to assess, given the suspected mode of action of perchlorate. However, the tests used in the Argus study were screening measures and would be unlikely to pick up subtle alterations in motor or cognitive function associated with moderate reductions in serum T_4 concentrations. Some important end points, such as auditory function and balance or coordination, were not assessed, so it is not surprising that no significant effects were observed in any of the behavioral measures except an increase in motor activity in male pups on one day of testing. Given the lack of sensitivity of the tests that were used, the committee concludes that the results are inadequate to determine whether or not gestational or lactational exposure to perchlorate affects behavioral function.

THYROID GLAND TUMORS

Few long-term cancer bioassays involving perchlorate exposure have been conducted.[2] In its draft risk assessment (EPA 2002a), EPA reviewed two studies that evaluated the occurrence of thyroid tumors in rodents exposed to perchlorate. First, Kessler and Kruskemper (1966), as cited in EPA (2002a), administered potassium perchlorate to male Wistar rats in drinking water at 0 or 1% for 24 months. EPA estimated the daily dose at 1,339 mg/kg. Histologic changes were similar to those produced by exposure to antithyroid agents. Four of the 11 treated rats developed benign

---

[2]Perchlorate has been evaluated in standard in vitro and in vivo assays used to assess genotoxicity (EPA 2002a). The results of those assays were negative.

tumors of the thyroid gland; no tumors were observed in the 20 control rats.

Second, Pajer and Kalisnik (1991) administered 0 or 1.2% sodium perchlorate in drinking water to female BALB/c mice for 46 weeks. There were three groups of controls and three groups of mice treated with perchlorate (12 mice per group). The estimated perchlorate dose was 2,147 mg/kg per day (EPA 2002a). One control group and one group of perchlorate-exposed mice were subjected to whole-body irradiation at 8 or 32 weeks with 0.8 Gy on 5 consecutive days at a dose rate of 1.45 Gy per minute for a total radiation dose of 4 Gy. Surviving animals were sacrificed at 46 weeks for histologic examination of the thyroid and pituitary glands. Thyroid follicular-cell carcinomas were observed in five of the six non-irradiated perchlorate-treated mice and in all 14 irradiated perchlorate-treated mice. No thyroid follicular-cell carcinomas were observed in the 22 control animals.

The committee found additional information in three other studies. First, Hiasa et al. (1987) found that ammonium perchlorate administered in the diet at 1,000 ppm was a thyroid gland tumor promoter in Wistar rats after tumor initiation with *N*-bis(2-hydroxypropyl)-nitrosamine (DHPN). After 19 weeks of treatment with perchlorate after DHPN administration, all 20 rats had thyroid adenomas, compared with one of 20 rats treated with DHPN alone. Rats treated with ammonium perchlorate alone had no thyroid adenomas.

Second, Fernandez Rodriguez et al. (1991) reported that female Wistar rats administered 1% potassium perchlorate in the drinking water for 1-12 months developed a progressive increase in thyroid gland weight and a diffuse hypertrophy and hyperplasia of follicular cells with increased vascularity and decreased lumenal colloid. After 6 months of treatment, multiple (often bilateral) follicular-cell nodules of complex morphology appeared in the diffusely enlarged thyroid glands with a follicular, papillary, or trabecular histologic pattern. The follicular cells comprising the nodules often had a more basophilic cytoplasm and dense nuclear chromatin than normal thyroid cells. The authors suggested that the thyroid nodules were probably due to overstimulation by TSH.

Third, Fernández-Santos et al. (2004) reported Ki-ras mutational analysis of thyroid follicular-cell lesions induced by the administration of radioactive iodine (50 microcuries ^{131}I) and potassium perchlorate (1% in drinking water to female Wistar rats) for 6, 12, and 18 months. No mutations were found in the amplified gene segment of any of the induced thyroid tumors. The results of the study suggested that Ki-ras activation via mutations at codons 12 and 13 is neither a constant nor an early event in the development of thyroid follicular-cell carcinoma in rats.

In a two-generation study (Argus 1999), thyroid follicular-cell adenomas were observed in three male rats. The following discussion focuses only on the male rats. On arrival, Charles River CR/CD rats were individually housed and assigned to groups (30 rats of each sex per group) at ammonium perchlorate doses of 0 (control), 0.3, 3.0, and 30 mg/kg per day in drinking water provided ad libitum. Male rats (P1 generation) were treated for at least 70 days before mating, through the mating period, and until sacrifice at the age of about 24-25 weeks. The male offspring from that mating (F1 generation) were treated in a similar manner from at least 70 days before their mating until sacrifice at the ages of about 21-22 weeks. The duration of dosing in the drinking water was about the same for the P1 and F1 groups of animals except that there was additional exposure of the F1 pups during the gestation and lactation periods. After sacrifice, many tissues, including the thyroid, of the P1 and F1 rats were examined histologically.

Slides from the Argus study were initially read by the Argus pathologist, and the findings are tabulated in the final report of the two-generation study (Argus 1999). Because of variability in the histologic criteria and examination among the various perchlorate studies, the slides from all the studies, including the two-generation study (thyroid gland only), were reread by an EPA pathologist and were reviewed by a pathology working group (Wolf 2000). The findings of the re-evaluation are listed in the tables prepared by EPA (Wolf 2001, Tables 14 through 21) and in the text of EPA's draft risk assessment (EPA 2002a). EPA's report of the re-evaluation states that two male rats (7094 and 7117) in the F1 generation treated at the high dose of perchlorate had follicular-cell adenomas and that one of them (7117) had two adenomas, for a total of three adenomas.

The original study pathology report indicates that there was also a follicular-cell adenoma in a control male rat in the P1 generation (3617). In the F1 generation, rat 7094 (F1 high-dose male) had a follicular-cell adenoma. For rat 7117 (F1 high-dose male), the original pathology report indicates that there was a focal proliferative lesion and that it was considered by the pathologist as hyperplasia, not neoplasia.

In view of the differences between the original Argus study pathology report and the EPA text and tabular information, the committee requested photomicrographs of the three rats. The committee confirmed that the control male rat in the P1 generation had a follicular-cell adenoma. The committee agreed that rat 7094 had a follicular-cell adenoma and that the lesion originally diagnosed by the Argus pathologist as focal hyperplasia (severe) in rat 7117 was a follicular-cell adenoma, which was also the conclusion of the review of the pathology working group. Thus, in the two-generation study, there were follicular-cell adenomas in one control male rat (3617) in the P1 generation and two high-dose male rats (7094 and 7117) in the F1 generation. There were no adenomas at lower doses or in the controls in the F1 generation.

The age at which the P1 and F1 generations were sacrificed and determined to have thyroid follicular-cell adenomas are comparable (about 5-6 months). No tumors were observed in the high-dose male rats in the P1 generation. The duration of treatment of the P1 generation was similar to that of the F1 generation, but the F1 generation may have had additional exposure during the gestation and lactation periods.

In its draft risk assessment, EPA considered the tumors in the F1 generation to be treatment-related and noted that they were particularly remarkable because they occurred at 19 weeks (Wolf 2000). EPA requested that the National Institute of Environmental Health Sciences perform statistical analyses of the tumor data in the F1 males, which involved an extensive Bayesian analysis that incorporated historical control data from F344 and Sprague-Dawley rats. It was concluded that thyroid adenomas were statistically increased in the high-dose (30-mg/kg) group of F1 animals sacrificed as adults (P2 generation) at 19 weeks and that the latency and incidence of the tumors were remarkable relative to the entirety of the National Toxicology Program database for this type of tumor in this strain of rat (Dunson 2001).

With respect to the observation of thyroid follicular-cell adenomas in the two-generation study, these would be expected in high-dose male rats in the presence of a markedly goitrogenic regimen, which existed under the conditions of the study. Although the tumor response observed in the high-dose F1 males is not statistically significant by conventional statistical analyses (Fisher's exact test), the tumors observed in the male high-dose group are most likely related to treatment. Spontaneous thyroid follicular-cell adenomas can occasionally be observed in control rats of this strain and age, so follicular-cell adenomas in rats treated with a markedly goitrogenic regimen would not be an unexpected or unusual finding.

The International Agency for Research on Cancer (IARC) held a working group on thyrotropic agents and issued a monograph in 2001 (IARC 2001). The monograph provided the following statements: "Agents that lead to the development of thyroid neoplasia through an adaptive physiological mechanism belong to a different category from those that lead to neoplasia through genotoxic mechanisms or through mechanisms involving pathological responses with necrosis and repair. Agents that induce thyroid follicular cell tumors in rodents by interfering with thyroid hormone homeostasis, can with some exceptions, notability the sulfonamides, also interfere with thyroid hormone homeostasis in humans if given at a sufficient dose for a sufficient time. These agents can be assumed not to be carcinogenic in humans at concentrations that do not lead to alterations in thyroid hormone homeostasis."

In addition, EPA's science policy document on the assessment of thyroid follicular-cell tumors notes that although there may be some qualitative similarities, there is evidence that "humans may not be as sensitive quantitatively to thyroid cancer development of thyroid-pituitary disruption as are rodents" (EPA 1998b). The increased sensitivity may be due to marked species differences in the physiology of the thyroid gland (EPA 1998a; Hill et al. 1989). The EPA and IARC documents provide guidance for the evaluation of thyroid follicular-cell tumors based on mode of action (for example, tumors secondary to hormone imbalance).

Thus, the committee reached the following two conclusions:

- In the case of perchlorate, follicular-cell tumors in rats are not an unexpected finding at doses that are goitrogenic.
- It is unlikely that perchlorate poses a risk of thyroid cancer in humans.

PBPK MODELING

Physiologically based pharmacokinetic (PBPK) modeling is one of the methods of choice for determining human equivalent exposures (HEEs) and adjusting default uncertainty factors associated with the derivation of reference doses and reference concentrations for lifetime human exposures from animal studies (EPA 2002a). Thus, EPA relied on a series of PBPK models developed by the Department of Defense after conducting a peer review to facilitate interspecies extrapolations in its draft perchlorate risk assessment (EPA 2002b, 2003). The PBPK models were initially developed to describe the disposition (absorption, distribution, metabolism, and elimination) of perchlorate in adult rats (Fisher 2000). As data became available and the mode of action for perchlorate-induced effects were shown to be mediated through interactions with iodide at the thyroid sodium (Na^+)/iodide (I^-) symporter (NIS), the initial model was expanded to include the disposition of iodide and the inhibition of iodide uptake at the NIS in pregnant rats and fetuses, lactating rats and neonates, and adult humans to address dose-response issues associated with potentially sensitive populations (Clewell et al. 2001, 2003a,b; Merrill 2001, Merrill et al. 2003).

It is currently impractical, and in many cases unethical, to validate

PBPK simulations of the kinetic or dynamic responses to chemical challenges in the human fetus or neonate. Therefore, the developers of the PBPK models proposed—and EPA concurred—to use a parallelogram approach to constrain predictions of equivalent exposures for human fetuses and infants corresponding to the no-observed-adverse-effect level (NOAEL) in animal studies (see Figure 4-4). According to this approach, an internal dose that is relevant to toxicity in the fetus or neonate (for example, the concentration of perchlorate in blood or serum or the inhibition of iodide uptake by the thyroid in the dams) is first determined for the NOAEL in the critical animal toxicity study. By applying factors for life-stage differences (such as pregnant or lactating female rat vs adult male rat) and for species differences (such as adult rat vs adult human or pregnant or lactating rat vs pregnant or lactating human), one can constrain PBPK simulations by using known physiologic measures and biochemical constants to estimate HEEs in potentially sensitive populations that may not be suitable for experimental validation.

The committee agrees with EPA that PBPK modeling constitutes the best available approach to determining HEEs and adjustments of default uncertainty factors when reference doses are based on animal data. The PBPK models developed by DOD for the adult rat, adult human, pregnant rat and fetus, and lactating rat and neonate represents the current state of the science for integrating available animal and human data on the disposition of perchlorate and iodide and the interactions between these anions at the level of the thyroid NIS. Although many of the PBPK model parameters had to be estimated on the basis of a small set of in vivo pharmacokinetic studies, enough studies were available for validation of model simulations over a broad range of doses of both perchlorate and iodide to lend confidence to the applicability of the models for extrapolating from animal-study doses to human exposures. Further details on the PBPK models developed for EPA's risk assessment can be found in Appendix E.

OTHER EFFECTS OF PERCHLORATE

Effects on Iodide Metabolism in Nonthyroid Tissues in Animals

The NIS is present in substantial amounts not only in the thyroid gland but also in several other tissues, including the salivary glands, mammary glands, stomach, choroid plexus of the brain, and ciliary body of the eye (see Chapter 2). Iodide that is transported into those tissues returns rapidly

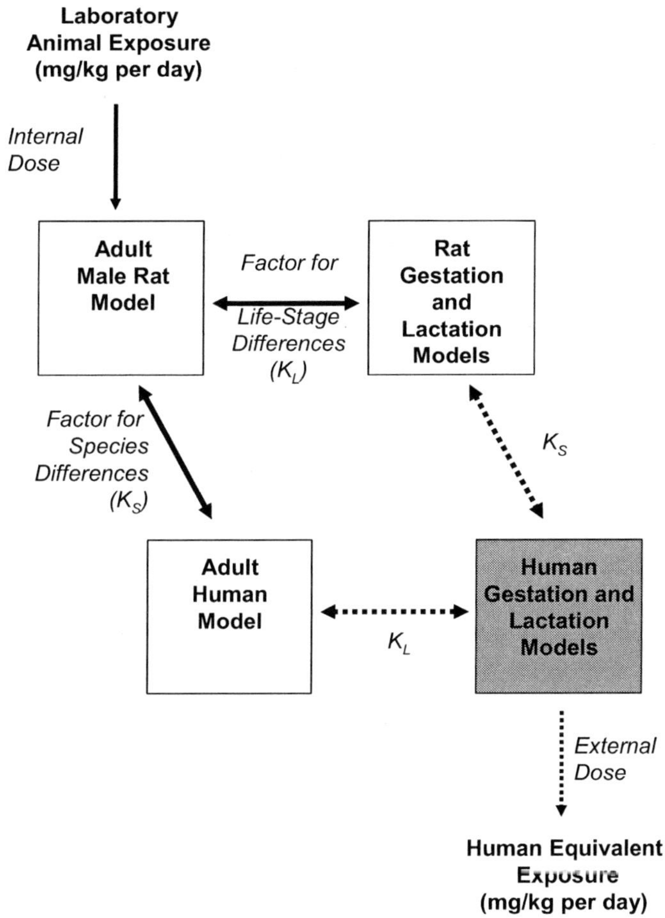

FIGURE 4-4 Parallelogram approach for using adult human, adult male rat, and female rat gestation and lactation models to estimate human equivalent exposures for human pregnancy and lactation models. Bold arrows indicate presence of validated PBPK models for estimating relationships between effective internal doses associated with key events and perchlorate exposure; dashed arrows and shaded box indicate theoretical extrapolations guided by physiologic and biochemical constraints in models and parallelogram approach.

to the circulation or is secreted into the saliva or breast milk. With molecular methods, very small amounts of NIS have also been detected in other tissues, including the heart, kidneys, lungs, and placenta, but there is little evidence that iodide is transported into them (Dohan et al. 2003). Three sites—mammary gland, placenta, and kidney—are briefly considered here because inhibition of NIS by perchlorate at these sites might contribute to in vivo effects of perchlorate.

The NIS of the mammary gland is important for two reasons: it can provide a means of transferring perchlorate to newborn infants, and perchlorate inhibition of the mammary gland NIS could decrease the iodide content of breast milk. Those issues were at least partially addressed by pharmacokinetic modeling of iodide and perchlorate metabolism in rats (Clewell et al. 2003b). Experiments conducted by Mahle et al. (2003) established that perchlorate is transferred from nursing rats to their pups.

Expression of NIS in the mammary gland differs from that in the thyroid gland. Mammary gland NIS is not affected by TSH, and the content of NIS in the mammary gland is low in virgin rats and high in lactating rats (Spitzweg et al. 1998; Tazebay et al. 2000). High NIS concentrations during lactation appear to be under the control of oxytocin and prolactin (Cho et al. 2000).

Iodide availability to the developing fetus is probably influenced by a variety of factors, including maternal iodide intake, placental and uterine deiodinating activity, and, potentially, placental NIS. Unlike the transport of iodide into thyroid and mammary tissue against a concentration gradient, placental transfer of iodide from mother to fetus is down a concentration gradient, so the importance of placental NIS in the process is not known. Nonetheless, inhibition of NIS in the placenta might reduce the transfer of iodide to the fetus, although it is not known to do so. In humans, expression of the NIS gene is higher in cultured placental cells from first-trimester pregnancies than in placental cells from third-trimester pregnancies, but NIS gene expression in crude placental extracts is similar at both times (Bidart et al. 2000). However, gene expression might not correlate with functional iodide transport, because NIS that is produced might not reach the cell membrane, where iodide transport occurs. In pregnant rats, NIS gene expression in the placenta was higher in response to iodide deficiency and potassium perchlorate administration on gestation day 21 than in control rats (Schroder-van der Elst et al. 2001).

Very small amounts of NIS are present in the cells that form the renal tubules. In the proximal tubules, it is primarily in the basolateral membrane of the cells (the side that is not exposed to urine) whereas it is more diffuse in the cells of the distal tubules (Spitzweg et al. 2001; Dohan et al. 2003). Little iodide is absorbed from the urine into the renal tubular cells, so it is unlikely that the small amounts of NIS in renal tubules reduce urinary iodide excretion. In a study in which some pregnant rats were fed perchlorate, their urinary excretion of iodide was higher than that of pregnant rats fed a normal diet, but the increase was due to decreased iodide uptake by the thyroid (Schröder-van der Elst et al. 2001).

General Toxicity

Despite its clinical use, few studies have focused on the general toxicity of perchlorate. Perchlorate has been widely used in biologic systems to study anion channels, anion-transport systems, and anion-constrained conformations of macromolecules. The literature was searched to identify potential interactions of perchlorate with non-NIS molecular targets. Such targets consisted of a variety of enzymes, such as cytochrome c and rabbit muscle enolase (Andersson et al. 1980; Robinson et al. 1983; Arai et al. 1984; Kornblatt et al. 1996). The concentrations of perchlorate required to bind to or inhibit those enzymes were up to several orders of magnitude greater than the concentrations needed to inhibit NIS. Likewise, perchlorate was found to affect the function of isolated cell systems, such as pancreatic islet cells and skeletal muscle fibers, but at much higher concentrations than those causing inhibition at the thyroid NIS (Gomolla et al. 1983; Sehlin 1987; Csernoch et al. 1987; Frankel and Sehlin 1994; Larsson-Nyren 1996; Larsson-Nyren et al. 2001). Thus, the high concentrations required for the interactions of perchlorate suggest that none of those other targets are important toxicologically in comparison with its more specific interaction with the NIS.

The primary publications that address a broad spectrum of toxicologic end points are reports of a 90-day study (Siglin et al. 2000) and of two reproductive and developmental studies (York et al. 2001a,b). The papers were important to the committee in determining whether any adverse effects, other than those associated with inhibition of iodide uptake by the thyroid, might arise from ingestion of perchlorate at low doses.

Siglin et al. (2000) administered ammonium perchlorate in drinking water to groups of Sprague-Dawley rats (10 rats of each sex per group) at 0, 0.01, 0.05, 0.2, 1.0, and 10 mg/kg per day for 14 or 90 days. Groups of rats also were evaluated 30 days after termination of the 90-day exposure at 0, 0.05, 1.0, and 10 mg/kg per day. In addition to thyroid function and thyroid histologic evaluations, standard toxicologic end points were evaluated, including clinical signs, body weights, food and water consumption, and routine hematologic and clinical-chemistry measures. An extensive list of tissues was examined microscopically in the control and high-dose rats; the liver, kidneys, lungs, and thyroids were examined in all animals.

Estrous cycles were monitored, and sperm were evaluated for count, motility, and structure. Bone samples also were evaluated for bone marrow micronucleus formation. Changes in thyroid hormones and TSH were observed as low as 0.01 mg/kg per day during exposure periods. However, changes in thyroid gland weight and thyroid histopathology were observed only at the highest dose (10 mg/kg per day). Although a few values were statistically significantly different from control values, no treatment-related effects were observed in any of the nonthyroid toxicologic measures. The authors concluded that the study provided further evidence that the thyroid is the primary target of perchlorate exposure in the rats.

York et al. (2001a) conducted a reproductive and developmental study (two-generation) in rats given ammonium perchlorate at daily doses of 0, 0.3, 3.0, and 30 mg/kg in drinking water. In addition to measures of thyroid pathology and function, organ weights and pathology of nonthyroid tissues were examined in the parental generation (P1) and two generations of pups (F1 and F2). Reproductive measures were examined in the P1 and F1 generations. There was some evidence of reduced sperm density, spermatid count, spermatid concentration, and spermatid density at the high dose administered, but the differences were not statistically significant. There were no other remarkable changes in any nonthyroid measure.

York et al. (2001b) conducted a developmental toxicity test in rabbits treated with ammonium perchlorate in drinking water from gestational days 6 through 28. Doses were adjusted to give targets of 0, 0.1, 1.0, 10, 30, and 100 mg/kg per day. No developmental anomalies were observed. Evidence of morphologic changes of thyroid follicular cells was observed at doses at 10 mg/kg per day or higher.

On the basis of the data reviewed, the committee concludes that perchlorate is very unlikely to have toxicologic effects at doses lower than those which would affect thyroid function.

Immunologic Effects

Perchlorate could theoretically produce several types of adverse immunologic reactions, including suppression of the function of a cellular component of the innate or adaptive immune systems, suppression of antibody- or cell-mediated responses, upregulation of immune-cell function or the full immune response, and induction of an immediate hypersensitivity (allergic) or delayed-type hypersensitivity reaction. The effects could possibly be mediated by a direct effect on the cells involved or indirectly by modulation of thyroid hormone-immune system homeostasis (Blalock 1994; Fabris et al. 1995; Klecha et al, 2000).

The potential of perchlorate to cause any of those adverse immunologic reactions has been studied in animals. In a 1993 study in which female rats were fed a low-iodide diet and perchlorate to induce iodide deficiency, the rats developed some of the pathologic changes of autoimmune thyroiditis and an increase in production of antithyroid antibodies (Mooij et al. 1993).

The tremendous expansion in recent years of knowledge of the components and functions of the human immune system has depended heavily on experiments in mice. Thus, mice are clearly the animals of choice to explore possible immunotoxic effects of perchlorate. However, it must be kept in mind that there are major species differences in various aspects of the immune system that preclude direct extrapolation of results from mice to humans (Mestas and Hughes 2004).

Among mouse strains, B6C3F1 female mice have become the test strain of choice for immunotoxicity studies and have been used in extensive studies of perchlorate by Keil et al. (1998, 1999). In those studies, mice were exposed to perchlorate in drinking water to achieve doses of 0.1, 1, 3, and 30 mg/kg per day. Mice were evaluated after 14 or 90 days of treatment and 30 days after 90 days of treatment. Serum T_4 concentrations were decreased significantly after 14 days in the 3- and 30-mg/kg groups and after 90 days in the 1-, 3-, and 30-mg/kg groups, but serum T_3 and TSH concentrations were normal.

The mice were evaluated with standard assays (Luster et al. 1988) for assessment of chemical-induced immunotoxicity. There were no significant or consistent differences between the control and perchlorate-exposed groups in any of the following outcomes: blood counts except reticulocytes only at the 90-day 3-mg/kg dose; weight or cellularity of the thymus, spleen, and bone marrow; CD4+/CD8+ lymphocyte counts; cytotoxic T-lymphocyte activity; IgG and IgM antibody responses to injections of sheep red blood cells (considered one of the most sensitive indicators of whether a chemical has immunosuppressive activity); macrophage nitric oxide production; resistance to a tumor challenge; and antinuclear antibody production (Keil et al. 1999). There was a persistent increase in NK cell activity only in the 30-mg/kg group. The one somewhat consistent finding was decreased phagocytosis of *Listeria monocytogenes* by peritoneal macrophages from perchlorate-exposed mice (all dose groups without a dose-response effect). The assay consisted of counting apparently internalized bacteria in stained smears. The biologic importance of the reduced phagocytosis was brought into question by in vivo experiments in which the perchlorate-exposed mice had normal resistance to infection by the same listeria.

In later studies (BRT-Burleson Research Technologies 2000 a,b,c), the antibody responses, as measured by the appearance of plaque-forming cells, to injections of sheep red blood cells were similar in normal and perchlorate-exposed mice. The local lymph node response to skin exposure to the sensitizing agent dinitrochlorobenzene was accentuated in the perchlorate groups, but the effect was not consistent, and there was no clear dose-response relationship. Perchlorate was not tested as a skin-sensitizing agent.

Those studies have been analyzed extensively by outside reviewers (RTI 1999; EPA 2002a,b; TERA 2002). The studies were generally judged to have been broad in scope and performed carefully with standard assays. The majority of reviewers concluded that the studies as a group had not demonstrated a causal relationship in mice between perchlorate ingestion in drinking water and any biologically meaningful stimulatory or inhibitory effect on the immune system. The committee agrees with previous reviewers and concludes that the evidence favors rejection of a causal relationship between ingestion of perchlorate and an immunotoxic effect in animals.

SUMMARY

The committee found that the animal studies of potential adverse effects of perchlorate provided qualitative information, but the usefulness of the studies for quantitatively estimating the risk of adverse effects in humans is small. The major conclusions from the animal data are summarized below.

- Perchlorate has an antithyroid effect on rats at high doses (30 mg/kg of ammonium perchlorate per day). That effect is characterized by decreases in serum thyroid hormone and increases in serum TSH with morphologic changes in the thyroid gland.
- The data are inadequate to determine whether or not a causal relationship exists between perchlorate exposure of pregnant rats and neurodevelopmental abnormalities in their pups, given the flaws in experimental design and methods in the studies conducted to evaluate that end point.
- The data are inadequate to determine whether or not perchlorate exposure during gestation and lactation in rats has effects on behavior, given the lack of sensitivity of the tests conducted to evaluate that end point.
- Exposure to perchlorate can increase the incidence of thyroid tumors in rats when the doses are high enough to decrease thyroid hormone production and increase TSH secretion.
- The data favor rejection of a causal relationship between perchlorate exposure and immunotoxicity.

- There are no data to suggest that perchlorate has effects that are not mediated through inhibition of iodide transport in the thyroid gland.
- It is not possible to extrapolate data quantitatively from rodents to humans for purposes of human health risk assessment. Most experimental studies in animals designed to characterize the effects of perchlorate exposure have been done in rats. However, rats are much more sensitive to agents that disturb thyroid function than are humans, so the relevance of rat studies in quantitative terms to humans is limited.

REFERENCES

Altman, J., and S.A. Bayer. 1997. Epilogue: Behavioral consequences of experimental interference with cerebellar development. Pp. 726-751 in Development of the Cerebellar System: In Relation to Its Evolution, Structure, and Functions. Boca Raton: CRC Press.

Alvarez-Dolado, M., M. Ruiz, J.A. Del Rio, S. Alcantara, F. Burgaya, M. Sheldon, K. Nakajima, J. Bernal, B.W. Howell, T. Curran, E. Soriano, and A. Munoz. 1999. Thyroid hormone regulates reelin and dab1 expression during brain development. J. Neurosci. 19(16):6979-6993.

Alvarez-Dolado, M., A. Cuadrado, C. Navarro-Yubero, P. Sonderegger, A.J. Furley, J. Bernal, and A. Munoz, A. 2000. Regulation of the L1 cell adhesion molecule by thyroid hormone in developing brain. Mol. Cell. Neurosci. 16(4):499-514.

Alvarez-Dolado, M., A. Figueroa, S. Kozlov, P. Sonderegger, A.J. Furley, and A. Munoz. 2001. Thyroid hormone regulates TAG-1 expression in the developing rat brain. Eur. J. Neurosci. 14(8):1209-1218.

Andersson, T., J. Angstrom, K.E. Falk, and S. Forsen. 1980. Perchlorate finding to cytochrome *c*: A magnetic and optical study. Eur. J. Biochem. 110(2):363-369.

Arai, Y., L. Oreland, and H. Kinemuchi. 1984. Effect of perchlorate treatment on mitochondrial MAO-A and -B activities. Med. Biol. 62(4):245-249.

Argus Research Laboratories, Inc. 1998. A Neurobehavioral Developmental Study of Ammonium Perchlorate Administered Orally in Drinking Water to Rats. ARGUS 1613-002. Argus Research Laboratories, Inc., Horsham, PA. [See also York, R.G., J. Barnett, W.R. Brown, R.H. Garman, D.R. Mattie, and D. Dodd. 2004. A rat neurodevelopmental evaluation of offspring, including evaluation of adult and neonatal thyroid, from mothers treated with ammonium perchlorate in drinking water. Int. J. Toxicol. 23(3):191-214.]

Argus Research Laboratories, Inc. 1999. Oral (Drinking Water) Two-Generation (One Litter per Generation) Reproduction Study of Ammonium Perchlorate in Rats. ARGUS 1416-001. Argus Research Laboratories, Inc., Horsham, PA.

Argus Research Laboratories, Inc. 2001. Hormone, Thyroid and Neurohistological Effects of Oral (Drinking Water) Exposure to Ammonium Perchlorate in Pregnant and Lactating Rats and in Fetuses and Nursing Pups Exposed to

Ammonium Perchlorate During Gestation or Via Maternal Milk. ARGUS 1416-003. Argus Research Laboratories, Inc., Horsham, PA.

Becker, J.T., J.A. Walker, and D.S. Olton. 1980. Neuroanatomical bases of spatial memory. Brain Res. 200(2):307-320.

Bekkedal, M.Y.V., T. Carpenter, J. Smith, C. Ademujohn, D. Maken, and D.R. Mattie. 2000. A Neurodevelopmental Study of Oral Ammonium Perchlorate Exposure on the Motor Activity of Pre-Weanling Rat Pup. Report No. TOXDET-00-03. Neurobehavioral Effects Laboratory, Naval Health Research Center Detachment (Toxicology), Wright-Patterson Air Force Base, OH.

Berbel, P., A. Guadano-Ferraz, M. Martinez, J.A. Quiles, R. Balboa, and G.M. Innocenti. 1993. Organization of auditory callosal connections in hypothyroid adult rats. Eur. J. Neurosci. 5(11):1465-1478.

Berry, S.M. 2000. Meta-analysis versus Large Trials: Resolving the controversy. Pp. 65-81 in Meta-Analysis in Medicine and Health Policy, D. Stangl, and D.A. Berry, eds. New York: Marcel Dekker.

Bianco, A.C., D. Salvatore, B. Gereben, M.J. Berry, and P.R. Larsen. 2002. Biochemistry, cellular and molecular biology, and physiological roles of the iodothyronine selenodeiodinases. Endocr. Rev. 23(1):38-89.

Bidart, J.M., L. Lacroix, D. Evain-Brion, B. Caillou, V. Lazar, R. Frydman, D. Bellet, S. Filetti, and M. Schlumberger. 2000. Expression of Na+/I- symporter and Pendred syndrome genes in trophoblast cells. J. Clin. Endocrinol. Metab. 85(11):4367-4372.

Blalock, J.E. 1994. The syntax of immune-neuroendocrine communication. Immunol. Today 15(11):504-511.

Bookstein, F.L., A.P. Streissguth, P.D. Sampson, P.D. Connor, and H.M. Barr. 2002. Corpus callosum shape and neuropsychological deficits in adult males with heavy fetal alcohol exposure. Neuroimage 15(1):233-251.

Brosvic, G.M., J.N. Taylor, and R.E. Dihoff. 2002. Influences of early thyroid hormone manipulations: Delays in pup motor and exploratory behavior are evident in adult operant performance. Physiol. Behav. 75(5):697-715.

BRT-Burleson Research Technologies, Inc. 2000a. Ammonium Perchlorate: Effect on Immune Function. Quality Assurance Audit: Study No. BRT 19990524 -- Plaque-Forming Cell (PFC) Assay; Study No. BRT 19990525 - Local Lymph Node Assay (LLNA) in Mice. BRT-Burleson Research Technologies, Inc., Raleigh, NC. June 30, 2000.

BRT-Burleson Research Technologies, Inc. 2000b. Addendum to Study Report: Ammonium Perchlorate: Effect on Immune Function. BRT 19990524 Study Protocol Plaque-Forming Cell (PFC) Assay; BRT 19990525 Study Protocol Local Lymph Node Assay (LLNA) in Mice. BRT-Burleson Research Technologies, Inc., Raleigh, NC. August 31, 2000.

BRT-Burleson Research Technologies, Inc. 2000c. Ammonium Perchlorate: Effect on Immune Function. Study Report. BRT 19990524 Study Protocol Plaque-Forming Cell (PFC) Assay; BRT 19990525 Study Protocol Local Lymph Node Assay (LLNA) in Mice. BRT-Burleson Research Technologies, Inc., Raleigh, NC.

Calvo, R., M.J. Obregon, C. Ruiz de Ona, B. Ferreiro, E. Escobar del Rey, and G. Morreale de Escobar. 1990. Thyroid hormone economy in pregnant rats near term: A "physiological" animal model of nonthyroidal illness? Endocrinology 127(1):10-16.

Cho, J.Y., R. Leveille, R. Kao, B. Rousset, A.F. Parlow, W.E. Burak Jr., E.L. Mazzaferri, and S.M. Jhiang. 2000. Hormonal regulation of radioiodide uptake activity and Na+/I- symporter expression in mammary glands. J. Clin. Endocrinol. Metab. 85(8):2936-2943.

Clewell, R.A., E.A. Merrill, and P.J. Robinson. 2001. The use of physiologically based models to integrate diverse data sets and reduce uncertainty in the prediction of perchlorate and iodide kinetics across life stages and species. Toxicol. Ind. Health 17(5-10):210-222.

Clewell, R.A., E.A. Merrill, K.O. Yu, D.A. Mahle, T.R. Sterner, D.R. Mattie, P.J. Robinson, J.W. Fisher, and J.M. Gearhart. 2003a. Predicting fetal perchlorate dose and inhibition of iodide kinetics during gestation: A physiologically-based pharmacokinetic analysis of perchlorate and iodide kinetics in the rat. Toxicol. Sci. 73(2):235-255.

Clewell, R.A., E.A. Merrill, K.O. Yu, D.A. Mahle, T.R. Sterner, J.W. Fisher, and J.M. Gearhart. 2003b. Predicting neonatal perchlorate dose and inhibition of iodide uptake in the rat during lactation using physiologically-based pharmacokinetic modeling. Toxicol. Sci. 74(2):416-436.

Crofton, K.M., D. Ding, R. Padich, M. Taylor, and D. Henderson. 2000. Hearing loss following exposure during development to polychlorinated biphenyls: A cochlear site of action. Hear. Res. 144(1-2):196-204.

Csernoch, L., L. Kovacs, and G. Szucs. 1987. Perchlorate and the relationship between charge movement and contractile activation in frog skeletal muscle fibers. J. Physiol. 390:213-227.

Dohan, O., A. De La Vieja, V. Paroder, C. Riedel, M. Artani, M. Reed, C.S. Ginter, and N. Carrasco. 2003. The sodium/iodide Symporter (NIS): Characterization, regulation, and medical significance. Endocr. Rev. 24(1):48-77.

Dohler, K.D., C.C. Wong, and A. von zur Muhlen. 1979. The rat as model for the study of drug effects on thyroid function: Consideration of methodological problems. Pharmacol. Ther. 5(1-3):305-318.

Dunson, D.B. 2001. Statistical Analysis of the Effects of Perchlorate on Neurobehavior (Motor Activity) in SD Rats. Memorandum to Annie M. Jarabek, National Center for Environmental Assessment (MD-52), U.S. Environmental Protection Agency, Research Triangle Park, NC, from David B. Dunson, Biostatistics Branch (MD A3-03), National Institute of Environmental Health Sciences, Research Triangle Park, NC. November 11, 2001. [Online]. Available: http://www.epa.gov/ncea/perchlorate/references2/ documents/100510.pdf [accessed July 19, 2004].

Elberger, A.J. 2003. Omaha Perchlorate Meeting – Module A. Analysis and Interpretation of Neurodevelopmental Rat Brain Morphometry Studies. Presentation at the Second Meeting on Assess the Health Implications of Perchlorate Ingestion, December 13, 2003, Irvine, CA.

EPA (U.S. Environmental Protection Agency). 1998a. Health Effects Test Guidelines OPPTS 870.6300 Developmental Neurotoxicity Study. EPA 712-C-98-239. Office of Prevention, Pesticides and Toxic Substances, U.S. Environmental Protection Agency, Washington, DC. [Online]. Available: http://www.epa.gov/opptsfrs/OPPTS_Harmonized/870_Health_Effects_Test_Guidelines/Series/870-6300.pdf [accessed August 23, 2004].

EPA (U.S. Environmental Protection Agency). 1998b. Assessment of Thyroid - Follicular Cell Tumors. EPA/630/R-97/002. Risk Assessment Forum, Office of Research and Development, U.S. Environmental Protection Agency, Washington, DC. [Online]. Available: http://cfpub.epa.gov/ncea/raf/recordisplay.cfm?deid=13102 [accessed November 22, 2004].

EPA (U.S. Environmental Protection Agency). 2002a. Perchlorate Environmental Contamination: Toxicological Review and Risk Characterization. External Review Draft. NCEA-1-0503. National Center for Environmental Assessment, Office of Research and Development, U.S. Environmental Protection Agency, Washington, DC. [Online]. Available: http://cfpub1.epa.gov/ncea/cfm/recordisplay.cfm?deid=24002 [accessed August 23, 2004]

EPA (U.S. Environmental Protection Agency). 2002b. Report on the Peer Review of the U.S. Environmental Protection Agency's Draft External Review Document "Perchlorate Environmental Contamination: Toxicological Review and Risk Characterization". EPA/635/R-02/003. National Center for Environmental Assessment, Office of Research and Development, U.S. Environmental Protection Agency, Washington, DC. [Online]. Available: http://www.epa.gov/ncea/pdfs/perchlorate/final_rpt.pdf [accessed August 23, 2004].

EPA (U.S. Environmental Protection Agency). 2003. Disposition of Comments and Recommendations for Revisions to "Perchlorate Environmental Contamination: Toxicological Review and Risk Characterization External Review Draft (January 16, 2002). Office of Research and Development, U.S. Environmental Protection Agency, Washington, DC. [Online]. Available: http://cfpub2.epa.gov/ncea/cfm/recordisplay.cfm?deid=72117 [accessed August 25, 2004].

Fabris, N., E. Mocchegiani, and M. Provinciali. 1995. Pituitary-thyroid axis and immune system: A reciprocal neuroendocrine-immune interaction. Horm. Res. 43(1-3):29-38.

Fernandez Rodriguez, A., H. Galera Davidson, M. Salguero Villadiego, A. Moreno Fernandez, I. Martin Lacave, and J. Fernandez Sanz. 1991. Induction of thyroid proliferative changes in rats treated with antithyroid compound. Anat. Histol. Embryol. 20(4):289-298.

Fernandez-Santos, J.M., M. De-Miguel, R. Gonzalez-Campora, M. Salguero-Villadiego, J.J. Cabrera, and H. Galera-Davidson. 2004. Ki-ras mutational analysis in rat follicular-cell proliferative lesions of the thyroid gland induced by radioactive iodine and potassium perchlorate. J. Endocrinol. Invest. 27(1):12-17.

Fisher, J.W. 2000. Consultative Letter, AFRL-HE-WP-CL-2000-0035, Physiological Model for Inhibition of Thyroidal Uptake of Iodide by Perchlorate in the Rat. Memorandum to Annie M. Jarabek, National Center for Environmental

Assessment (MD-52), U.S. Protection Agency, Research Triangle Park, NC, from Jeffrey W. Fisher, Air Force Research Laboratory, Wright-Patterson Air Force Base, OH. June 28, 2000.

Frankel, B.J., and J. Sehlin. 1994. Effect of perchlorate on glucose–stimulated insulin release and $45Ca^{2+}$ uptake in pancreatic islets from diabetic Chinese hamsters. Pancreas 9(5):550-557.

Fukuda, H., K. Ohshima, M. Mori, I. Kobayashi, and M.A. Greer. 1980. Sequential changes in the pituitary-thyroid axis during pregnancy and lactation in the rat. Endocrinology 107(6):1711-1716.

Gauger, K.J., Y. Kato, K. Haraguchi, H.J. Lehmler, L.W. Robertson, R. Bansal, and R. Zoeller. 2004. Polychlorinated biphenyls (PCBs) exert thyroid hormone-like effects in the fetal rat brain but do not bind to thyroid hormone receptors. Environ. Health Perspect. 112(5):516-523.

Geller, A.M. 2003. Revised Brain Morphometry Analysis Incorporating Consultant in Veterinary Pathology (2003) Review of Morphometry Data from Argus 1416-003. Memorandum to Annie M. Jarabek, National Center for Environmental Assessment, U.S. Environmental Protection Agency, Research Triangle Park, NC, from A.M. Geller, National Health and Environmental Effects Research Laboratory, U.S. Environmental Protection Agency, Research Triangle Park, NC. September 19, 2003.

Glinoer, D. 1997. The regulation of thyroid function in pregnancy: Pathways of endocrine adaptation from physiology to pathology. Endocr. Rev. 18(3):404-433.

Goldey, E.S., and K.M. Crofton. 1998. Thyroxine replacement attenuates hypothyroxinemia, hearing loss, and motor deficits following developmental exposure to Aroclor 1254 in rats. Toxicol. Sci. 45(1):94-105.

Goldey, E.S., L.S. Kehn, C. Lau, G.L. Rehnberg, and K.M. Crofton. 1995a. Developmental exposure to polychlorinated biphenyls (Aroclor 1254) reduces circulating thyroid hormone concentrations and causes hearing deficits in rats. Toxicol. Appl. Pharmacol. 135(1):77-88.

Goldey, E.S., L.S. Kehn, G.L. Rehnberg, and K.M. Crofton. 1995b. Effects of developmental hypothyroidism on auditory and motor function in the rat. Toxicol. Appl. Pharmacol. 135(1):67-76.

Gomolla, M., G. Gottschalk, and H.C. Luttgau. 1983. Perchlorate-induced alterations in electrical and mechanical parameters of frog muscle fibres. J. Physiol. 343:197-214.

Gould, E., A. Westlind-Danielsson, M. Frankfurt, and B.S. McEwen. 1990. Sex differences and thyroid hormone sensitivity of hippocampal pyramidal cells. J. Neurosci. 10(3):996-1003.

Gravel, C., and R. Hawkes. 1990. Maturation of the corpus callosum of the rat: I. Influence of thyroid hormones on the topography of callosal projections. J. Comp. Neurol. 291(1):128-146.

Greer, M.A., G. Goodman, R.C. Pleus, and S.E. Greer. 2002. Health effects assessment for environmental perchlorate contamination: The dose response for inhibition of thyroidal radioiodine uptake in humans. Environ. Health Perspect. 110(9):927-937.

Hard, G.C. 1998. Recent developments in the investigation of thyroid regulation and thyroid carcinogenesis. Environ. Health Perspect. 106(8):427-436.

Harry, J. 2001. Re: Comments on Original Experimental Design, Study Performance, and Brain Morphometry Results of Argus Research Laboratories, Inc. 14 March 2001 Study (Protocol Number 1416-003) and Supplemental Materials Provided by Dr. Robert Garman, Consultants in Veterinary Pathology, Inc. Letter to Annie M. Jarabek, National Center for Environmental Assessment, U.S. Environmental Protection Agency, from Jean Harry, Acting Chief, Laboratory of Toxicology Neurotoxicology Group Leader, National Institute of Environmental Health Sciences, Research Triangle Park, NC. October 11, 2001.

Hiasa, Y., Y. Kitahori, Y. Kato, M. Ohshima, N. Konishi, T. Shimoyama, Y. Sakaguchi, H. Hashimoto, S. Minami, and Y. Murata. 1987. Potassium perchlorate, potassium iodide, and propylthiouracil: Promoting effect on the development of thyroid tumors in rats treated with N-bis(2-hydroxypropyl)-nitrosamine. Jpn. J. Cancer Res. 78(12):1335-1340.

Hill, R.N., L.S. Erdreich, O.E. Paynter, P.A. Roberts, S.L. Rosenthal, and C.F. Wilkinson. 1989. Thyroid follicular cell carcinogenesis. Fundam. Appl. Toxicol. 12(4):629-697.

IARC (International Agency for Research on Cancer). 2001. Pp. 40-41 in Some Thyrotropic Agents, IARC Monographs on the Evaluation of Carcinogenic Risk to Humans, Vol. 79. Lyon, France: International Agency for Research on Cancer.

Keil, D., A. Warren, M. Jenny, J. EuDaly, and R. Dillard. 1998. Effects of Ammonium Perchlorate on Immunotoxicological, Hematological, and Thyroid Parameters in B6C3F1 Female Mice. DSWA01-97-0008. Medical University of South Carolina, Charleston, SC. September 30, 1998. [Online]. Available: http://www.epa.gov/ncea/perchlorate/references2/documents/44966.pdf [accessed August 25, 2004.]

Keil, D., A. Warren, M. Jenny, J. EuDaly, and R. Dillard. 1999. Effects of Ammonium Perchlorate on Immunotoxicological, Hematological, and Thyroid Parameters in B6C3F1 Female Mice, Final Report. DSWA01-97-0008. Medical University of South Carolina, Charleston, SC. June 19, 1999. [Online]. Available: http://www.epa.gov/ncea/perchlorate/references2/documents/99555.pdf [accessed July 16, 2004].

Kessler, M.E., and H.L. Kruskemper. 1966. Experimental thyroid tumors caused by long-term potassium perchlorate administration. [in German]. Klin Wochenschr. 44(19):1154-1156.

Klecha, A.J., A.M. Genaro, A.E. Lysionek, R.A. Caro, A.G. Coluccia, and G.A. Cremaschi. 2000. Experimental evidence pointing to the bidirectional interaction between the immune system and the thyroid axis. Int. J. Immunopharmacol. 22(7):491-500.

Klintsova, A.Y., C.R. Goodlett, and W.T. Greenough. 2000. Therapeutic motor training ameliorates cerebellar effects of postnatal binge alcohol. Neurotoxicol. Teratol. 22(1):125-132.

Knipper, M., C. Zinn, H. Maier, M. Praetorius, K. Rohbock, I. Kopschall, and U. Zimmermann. 2000. Thyroid hormone deficiency before the onset of hearing causes irreversible damage to peripheral and central auditory systems. J. Neurophysiol. 83(5):3101-3112.

Koibuchi, N., S. Yamaoka, and W.W. Chin. 2001. Effect of altered thyroid status on neurotrophin gene expression during postnatal development of the mouse cerebellum. Thyroid 11(3):205-210.

Kornblatt, M.J., A. Al-Ghanim, and J.A. Kornblatt. 1996. The effects of sodium perchlorate on rabbit muscle enolase--Spectral characterization of the monomer. Eur. J. Biochem. 236(1):78-84.

Larsson-Nyren, G. 1996. Perchlorate is hypoglycemic by amplifying glucose-stimulated insulin secretion in mice. Acta Physiol. Scand. 158(1):71-76.

Larsson-Nyren, G., J. Sehlin, P. Rorsman, and E. Renstrom. 2001. Perchlorate stimulates insulin secretion by shifting the gating of L-type $Ca^{2}+$ currents in mouse pancreatic B-cells towards negative potentials. Pflugers Arch. 441(5):587-595.

Lasky, R.E., J.J. Widholm, K.M. Crofton, and S.L. Schantz. 2002. Perinatal exposure to Aroclor 1254 impairs distortion product otoacoustic emissions (DPOAEs) in rats. Toxicol. Sci. 68(2):458-464.

Lauder, J.M. 1977. The effects of early hypo- and hyperthyroidism on the development of rat cerebellar cortex. III. Kinetics of cell proliferation in the external granular layer. Brain Res. 126(1):31-51.

Lawrence, J.E., S.H. Lamm, S. Pino, K. Richman, and L.E. Braverman. 2000. The effect of short-term low-dose perchlorate on various aspects of thyroid function. Thyroid 10(8):659-663.

Legrand, J. 1986. Thyroid hormone effects on growth and development. Pp. 503-534 in Thyroid Hormone Metabolism, G. Hennemann, ed. New York: Marcel Dekker.

Luster, M.I., A.E. Munson, P.T. Thomas, M.P. Holsapple, J.D. Fenters, K.L. White, Jr., L.D. Lauer, D.R. Germolec, G.J. Rosenthal, and J.H. Dean. 1988. Development of a testing battery to assess chemical-induced immunotoxicity: National Toxicology Program's guidelines for immunotoxicity evaluation in mice. Fundam. Appl. Toxicol. 10(1):2-19.

Magara, F., L. Ricceri, D.P. Wolfer, and H.P. Lipp. 2000. The acallosal mouse strain I/LnJ: A putative model of ADHD? Neurosci. Biobehav. Rev. 24(1):45-50.

Mahle, D.A., K.O. Yu, L. Narayanan, D.R. Mattie, and J.W. Fisher. 2003. Changes in cross-fostered Sprague-Dawley rat litters exposed to perchlorate. Int. J. Toxicol. 22(2):87-94.

McClain, R.M. 1995. Mechanistic considerations for the relevance of animal data on thyroid neoplasia to human risk assessment. Mutat. Res. 333(1-2):131-142.

Merrill, E.A. 2001. Consultative Letter AFRL-HE-WP-CL-2001-0008. PBPK Model for Perchlorate-induced Inhibition of Radioiodide Uptake in Humans. Memorandum with attachments to A. Jarabek, National Center for Environmental Assessment, Washington, DC, from E.A. Merrill, Air Force Research Laboratory, Wright-Patterson AF, OH. June 5, 2001.

Merrill, E.A., R.A. Clewell, J.M. Gearhart, P.J. Robinson, T.R. Sterner, K.O. Yu, D.R. Mattie, and J.W. Fisher. 2003. PBPK predictions of perchlorate distribution and its effect on thyroid uptake of radioiodide in the male rat. Toxicol. Sci. 73(2):256-269.

Mestas, J., and C.C. Hughes. 2004. Of mice and not men: Differences between mouse and human immunology. J. Immunol. 172(5):2731-2738.

Mooij, P., H.J. de Wit, A.M. Bloot, M.M. Wilders-Truschnig, and H.A. Drexhage. 1993. Iodine deficiency induces thyroid autoimmune reactivity in Wistar rats. Endrocrinology 133(3):1197-1204.

Ng, L., R.J. Goodyear, C.A. Woods, M.J. Schneider, E. Diamond, G.P. Richardson, M.W. Kelley, D.L. Germain, V.A. Galton, and D. Forrest. 2004. Hearing loss and retarded cochlear development in mice lacking type 2 iodothyronine deiodinase. Proc. Natl. Acad. Sci. U.S.A. 101(10):3474-3479.

Obregon, M.J., C. Ruiz de Ona, R. Calvo, F. Escobar del Rey, and G. Morreale de Escobar. 1991. Outer ring iodothyronine deiodinases and thyroid hormone economy: Responses to iodine deficiency in the rat fetus and neonate. Endocrinology 129(5):2663-2673.

Olton, D.S., J.T. Becker, and G.E. Handelmann. 1979. Hippocampus, space and memory. Behav. Brain Sci. 2:313-365.

Pajer, Z., and M. Kalisnik. 1991. The effect of sodium perchlorate and ionizing irradiation on the thyroid parenchymal and pituitary thyrotropin cells. Oncology 48(4):317-320.

Rami, A., A.J. Patel, and A. Rabie. 1986. Thyroid hormone and development of the rat hippocampus: Morphological alterations in granule and pyramidal cells. Neuroscience 19(4):1217-1226.

Ribak, C.E. 1986. Contemporary methods in neurocytology and their application to the study of epilepsy. Adv. Neurol. 44:739-764.

Robinson Jr., J.B., J.M. Strottmann, and E. Stellwagen. 1983. A globular high spin form of ferricytochrome c. J. Biol. Chem. 258(11):6772-6776.

Roegge, C.S., V.C. Wang, B.E. Powers, A.Y. Klintsova, S. Villareal, W.T. Greenough, and S.L. Schantz. 2004. Motor impairment in rats exposed to PCBs and methylmercury during early development. Toxicol. Sci. 77(2):315-324.

Rovet, J.F. 2002. Congenital hypothyroidism: An analysis of persisting deficits and associated factors. Neuropsychol. Dev. Cogn. Sect. C. Child Neuropsychol. 8(3):150-162.

RTI (Research Triangle Institute). 1999. Pp. 3-28 to 3-31 in Perchlorate Peer Review Workshop Report. EPA Contract No. 68-W98-085. RTI No. 7200-019. Prepared for Office of Solid Waste, U.S. Environmental Protection Agency, Washington, DC, by Center for Environmental Analysis, Research

Triangle Institute, Research Triangle Park, NC. May 13, 1999. [Online]. Available: http://www.epa.gov/ncea/perchlorate/references2/documents/98020.pdf [accessed August 25, 2004.]

Schröder-van der Elst, J.P., D. van der Heide, J. Kastelijn, B. Rousset, and M.J. Obregón. 2001. The expression of the sodium/iodide symporter is up-regulated in the thyroid of fetuses of iodine-deficient rats. Endocrinology 142(9):3736-3741.

Schwegler, H., W.E. Crusio, and I. Brust. 1990. Hippocampal mossy fibers and radial-maze learning in the mouse: A correlation with spatial working memory but not with non-spatial reference memory. Neuroscience 34(2):293-298.

Seeger, G., U. Gartner, M. Holzer, and T. Arendt. 2003. Constitutive expression of p21H-Ras(Val12) in neurons induces increased axonal size and dendritic microtubule density in vivo. J. Neurosci. Res. 74(6):868-874.

Sehlin, J. 1987. Effect of perchlorate on calcium uptake and insulin secretion in mouse pancreatic islets. Biochem. J. 248(1):109-115.

Siglin, J.C., D.R. Mattie, D.E. Dodd, P.K. Hildebrandt, and W.H. Baker. 2000. A 90-day drinking water toxicity study in rats of the environmental contaminant ammonium perchlorate. Toxicol. Sci. 57(1):61-74.

Spitzweg, C., W. Joba, W. Eisenmenger, and A.E. Heufelder. 1998. Analysis of human sodium iodide symporter gene expression in extrathyroidal tissues and cloning of its complementary deoxyribonucleic acids from salivary gland, mammary gland, and gastric mucosa. J. Clin. Endocrinol. Metab. 83(5):1746-1751.

Spitzweg, C., C.M. Dutton, M.R. Castro, E.R. Bergert, J.R. Goellner, A.E. Heufelder, and J.C. Morris. 2001. Expression of the sodium iodide symporter in human kidney. Kidney Int. 59(3):1013-1023.

Steinmetz, J.E. 1996. The brain substrates of classical eyeblink conditioning in rats. Pp. 89-114 in The Acquisition of Motor Behavior in Vertebrates, J.R. Bloedel, T.J. Ebner, and S.P. Wise, eds. Cambridge, MA: MIT Press.

Tazebay, U.H., I.L. Wapnir, O. Levy, O. Dohan, L.S. Zuckier, Q.H. Zhao, H.F. Deng, P.S. Amenta, S. Fineberg, R.G. Pestell, and N. Carrasco. 2000. The mammary gland iodide transporter is expressed during lactation and in breast cancer. Nat. Med. 6(8):871-878.

TERA(Toxicology Excellence for Risk Assessment). 2001. Report on Five Expert Reviews of the Primedica 2001 Study Report (Hormone, Thyroid and Neurohistological Effects of Oral (Drinking Water) Exposure to Ammonium Perchlorate in Pregnant and Lactating Rats and in Fetuses and Nursing Pups Exposed to Ammonium Perchlorate During Gestation or Via Maternal Milk, March 2001). Prepared for Perchlorate Study Group. May 18, 2001.

TERA (Toxicology Excellence for Risk Assessment). 2002. Pp. 6-7 in Quantitative Evaluation of Perchlorate Risk Assessment, February 2002. Toxicology Excellence for Risk Assessment, Cincinnati, OH. [Online]. Available: http://www.tera.org/Perchlorate/complete/quantitative%20eval%20perchlorate.pdf [accessed July 16, 2004.].

Versloot, P.M., J. Gerritsen, L. Boogerd, J.P. Schröder-van der Elst, and D. van der Heide. 1994. Thyroxine and 3,5,3'-triiodothyronine production, metabolism, and distribution in pregnant rat near term. Am. J. Physiol. 267(6 Pt 1):E860-E867.

Vranckx, R., M. Rouaze-Romet, L. Savu, P. Mechighel, M. Maya, and E.A. Nunez. 1994. Regulation of rat thyroxine-binding globulin and transthyretin: Studies in thyroidectomized and hypophysectomized rats given tri-iodothyronine or/and growth hormone. J. Endocrinol. 142(1):77-84.

Wahlsten, D. 2002. Perchlorate Effects on Rat Brain Morphometry: A Critical Evaluation. Submitted to Eastern Research Group, Inc. for the U.S. EPA /ORD Peer Review Workshop-Perchlorate Environmental Contamination: Toxicological Review and Risk Characterization, March 5-6, Sacramento, CA. February 19. [Online]. Available: http://www.perchloratesymposium.com/docs/Wahlsten2002.pdf [accessed August 23, 2004].

Wasniewska, M., F. De Luca, S. Siclari, G. Salzano, M.F. Messina, F. Lombardo, M. Valenzise, C. Ruggeri, and T. Arrigo. 2002. Hearing loss in congenital hypothalamic hypothyroidism: A wide therapeutic window. Hear Res. 172(1-2):87-91.

Widholm, J.J., B.W. Seo, B.J. Strupp, R.F. Seegal, and S.L. Schantz. 2003. Effects of perinatal exposure to 2,3,7,8-tetrachlorodibenzo-p-dioxin on spatial and visual reversal learning in rats. Neurotoxicol. Teratol. 25(4):459-471.

Wolf, D.C. 2000. Report of the Peer Review of the Thyroid Histopathology From Rodents and Rabbits Exposed to Ammonium Perchlorate in the Drinking Water. Memorandum to Annie Jarabek and William Farland, National Center for Environmental Assessment, U.S. Environmental Protection Agency, Research Triangle Park, NC, from Douglas C. Wolf, Environmental Carcinogenesis Division, National Health and Environmental Effects Research Laboratory, Research Triangle Park, NC. May 5. 2000. [Online]. Available: http://www.epa.gov/ncea/pdf/perchlorate/ea121000.pdf [accessed August 26, 2004].

Wolf, D.C. 2001. Erratum to the Report of the Peer Review of the Thyroid Histopathology From Rodents and Rabbits Exposed to Ammonium Perchlorate in Drinking Water. Memorandum to Annie Jarabek and William Farland, National Center for Environmental Assessment, U.S. Environmental Protection Agency, Research Triangle Park, NC, from Douglas C. Wolf, Environmental Carcinogenesis Division, National Health and Environmental Effects Research Laboratory, Research Triangle Park, NC. October 26, 2001.

York, R.G., W.R. Brown, M.F. Girard, and J.S. Dollarhide. 2001a. Two-generation reproduction study of ammonium perchlorate in drinking water in rats evaluates thyroid toxicity. Int. J. Toxicol. 20(4):183-197.

York, R.G., W.R. Brown, M.R. Girard, and J.S. Dollarhide. 2001b. Oral (drinking water) developmental toxicity study of ammonium perchlorate in New Zealand white rabbits. Int. J. Toxicol. 20(4):199-205.

Yu, K.O., L. Narayanan, D.R. Mattie, R.J. Godfrey, P.N. Todd, T.R. Sterner, D.A.

Mahle, M.H. Lumpkin, and J.W. Fisher. 2002. The pharmacokinetics of perchlorate and its effect on the hypothalamus-pituitary-thyroid axis in the male rat. Toxicol. Appl. Pharmacol. 182(2):148-159.

Zoeller, T.R., A.L. Dowling, C.T. Herzig, E.A. Iannacone, K.J. Gauger, and R. Bansal. 2002. Thyroid hormone, brain development, and the environment. Environ. Health Perspect. 110(Suppl. 3):355-361.

5

Risk Characterization of Perchlorate

THE committee was charged with reviewing the relevant data on the health effects of perchlorate and the findings in the 2002 U.S. Environmental Protection Agency (EPA) report *Perchlorate Environmental Contamination: Toxicological Review and Risk Characterization.* Chapters 2, 3, and 4 of the present report contain the committee's review of the key studies, including those discussed in EPA's draft risk assessment, and the committee's findings will not be repeated here. This chapter provides the committee's assessment of the mode-of-action model of perchlorate toxicity, the definition of adverse effect, the point of departure, and the use of uncertainty factors to derive a reference dose (RfD) for daily oral exposures to perchlorate.

MODE-OF-ACTION MODEL

EPA's proposed mode-of-action model of perchlorate toxicity is provided in Chapter 1 (see Figure 1-2). EPA's model represents a continuum of possible health effects of perchlorate exposure and predicts that perchlorate exposure leads to inhibition of iodide uptake by the thyroid, which causes decreases in thyroxine (T_4) and triiodothyronine (T_3) production and then increases in thyrotropin (thyroid-stimulating hormone, TSH) secretion. At that point, EPA's model diverges into two pathways, one labeled as children's health risk and the other labeled as human health risk. EPA's model predicts that the changes in thyroid hormone production and increases in TSH production could lead to altered development and ultimately birth defects in children and to thyroid hyperplasia and ultimately thyroid tumors in adults.

The committee thinks that EPA's mode-of-action model adequately represents the possible early sequence of events after perchlorate exposure, but it does not think that the model provides an accurate representation of events that follow changes in thyroid hormone and TSH production. Specifically, the development of thyroid tumors as an ultimate result of perchlorate exposure is an unlikely outcome in humans. As discussed in Chapter 4, the committee is not surprised that rats treated with moderate or high doses of perchlorate would develop thyroid follicular-cell tumors. Rats are sensitive to the development of thyroid tumors because their thyroid function is easily disrupted. Humans are much less susceptible than rats to disruption of thyroid function and therefore are not likely to develop thyroid tumors as a result of perchlorate exposure. The committee concludes that the most reasonable pathway of events after changes in thyroid hormone and TSH secretion would be thyroid hypertrophy or hyperplasia, possibly leading to hypothyroidism. At that point, the pathway would diverge to two potential outcomes: (1) metabolic sequelae, such as decreased metabolic rate and slowing of the function of many organ systems, occurring at any age, and (2) abnormal growth and development in fetuses and children. The committee's suggested mode-of-action model is provided in Figure 5-1.

The committee emphasizes that inhibition of iodide uptake by the thyroid has been the only consistently documented effect of perchlorate exposure in humans. The continuum of possible effects of iodide-uptake inhibition caused by perchlorate exposure is only proposed and has not been demonstrated in humans exposed to perchlorate (with the exception that in patients with hyperthyroidism doses of 200 mg daily or higher may reduce thyroid secretion). More important, the outcomes at the end of the continuum are not inevitable consequences of perchlorate exposure. As discussed in Chapter 2, the body can compensate for decreases in T_4 and T_3 production unless there is a severe pre-existing thyroid disease. Specifically, the resulting increase in TSH secretion can return T_4 and T_3 production to normal without causing adverse effects on human health.

ADVERSE EFFECT AND KEY EVENT

EPA defines changes in serum thyroid hormone and TSH concentrations as adverse effects. The effects that would be downstream of those changes in its mode-of-action model would also be considered adverse effects. EPA states that the neurodevelopmental and neoplastic outcomes "confirm that the perturbation of the thyroid hormone economy should be

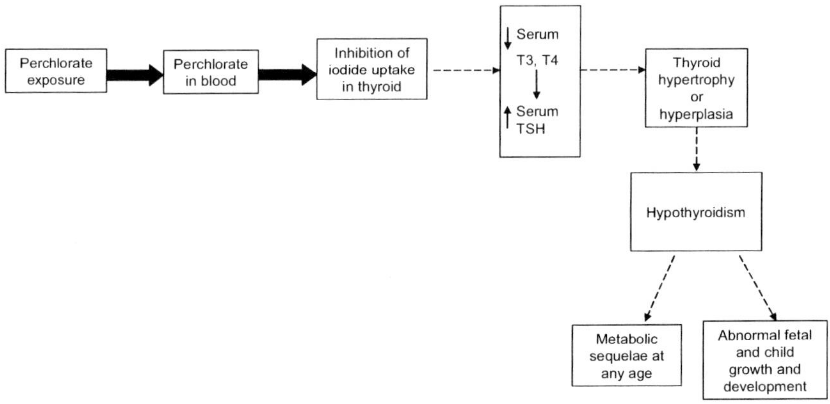

FIGURE 5-1 Committee's suggested mode-of-action model of perchlorate toxicity in humans. Solid arrows represent outcomes that have been observed in humans during perchlorate exposure. Dashed arrows represent outcomes that have not been clearly demonstrated in humans exposed to perchlorate but that are biologically possible in the absence of adequate compensation. The thyroid response to increased serum TSH and an independent increase in thyroid iodide uptake would raise T_3 and T_4 production to normal and therefore usually prevent the later steps from occurring.

viewed as adverse" (see EPA 2002a, p. 7-10). The committee, however, does not view transient changes in serum thyroid hormone and TSH concentrations as adverse health effects; it considers them to be biochemical changes that could precede adverse effects. Given its mode-of-action model, the committee concludes that hypothyroidism is the first adverse effect in the mode-of-action model (see Figure 5-2). Any effects downstream of hypothyroidism clearly would be adverse.

EPA developed its risk assessment by using data on effects that it views as adverse. However, the committee does not think that hypothyroidism—the effect that the committee views as adverse—should be used as the basis of a perchlorate risk assessment. It recommends that the key biochemical event be used as the basis of the perchlorate risk assessment. EPA and the committee agree that the key event in the continuum of possible effects of perchlorate exposure is the inhibition of iodide uptake by the thyroid. It is the obligatory initial step in the continuum of possible effects of perchlorate exposure, and thyroid uptake of iodide (as radioiodide) can be measured easily and reliably. *Inhibition of iodide uptake by the thyroid clearly is not an adverse effect; however, if it does not occur, there is no progression to*

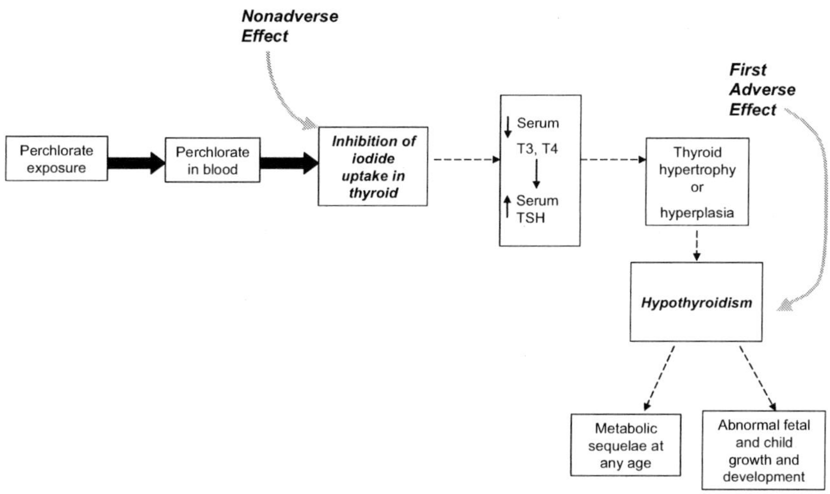

FIGURE 5-2 Committee's suggested mode-of-action model for perchlorate toxicity in humans indicating first adverse effect in the continuum.

adverse health effects (see Figure 5-2). The committee views its recommendation to use inhibition of iodide uptake by the thyroid as the basis of the perchlorate risk assessment to be the most health-protective and scientifically valid approach.

POINT OF DEPARTURE

A primary purpose of EPA's perchlorate risk assessment was to calculate an oral RfD. EPA (2002b) currently defines the RfD as

> an estimate (with uncertainty spanning perhaps an order of magnitude) of a daily oral exposure to the human population (including sensitive subgroups) that is likely to be without risk of deleterious effects during a lifetime. It can be derived from a NOAEL [no-observed-adverse-effect level], LOAEL [lowest-observed-adverse-effect level], or BMD [benchmark dose], with UFs [uncertainty factors] generally applied to reflect limitations of the data used.

The RfD definition uses several terms that should be defined. The NOAEL is the highest dose at which no *adverse* health effects have been observed, and the LOAEL is the lowest dose at which *adverse* health effects have been observed (EPA 2000). The NOAEL is often confused with the no-observed-effect level (NOEL), and a clear distinction should be made between the two terms. The NOEL is the highest dose "at which there are no statistically or biologically significant increases in the frequency or severity of any effect between the exposed population and its appropriate control" (ITER 2004). Thus, a NOAEL is based on an *adverse* effect, and a NOEL is based on a *nonadverse* effect.

Traditionally, NOAELs and LOAELs have been used to derive RfDs (EPA 2004). More recently, BMDs (or their lower confidence limits) calculated from mathematical modeling of dose-response data have been used to derive RfDs. Use of the BMD method has increased because it is seen as a more quantitative approach that accounts for variability in observed responses over an entire dose range and incorporates uncertainty due to characteristics of study design (EPA 2000). However, the vast majority of RfDs are based on NOAELs and LOAELs from the literature (EPA 2004).

The first step in deriving an RfD is a comprehensive review of all relevant human and animal data (EPA 2004). Traditionally, a critical effect and a critical study are then identified that serve as the point of departure for the risk assessment. Human or animal data can be used, but human data are preferred when sufficient data are available (EPA 2002b). Typically, a NOAEL or LOAEL is identified from the critical study on which the RfD can be based. As noted above, mathematical modeling of the dose-response data in the study can also provide a BMD on which the RfD can be based. As a final step in the RfD process, uncertainty factors are applied to the NOAEL, LOAEL, or BMD to extrapolate from the study population to the general human population, which includes sensitive groups. The individual uncertainty factors used to derive an RfD are discussed in the following sections.

For the perchlorate risk assessment, EPA based its point of departure on reported changes in brain morphometry, thyroid histopathology, and serum thyroid hormone concentrations after oral administration of perchlorate to rats. For several reasons, the committee does not think that the animal data or the outcomes selected by EPA should be used as the basis of the perchlorate risk assessment. As discussed in Chapter 4, the rat is a good quantitative model for assessing inhibition of iodide uptake by the thyroid caused by perchlorate exposure, but it is only a good *qualitative* model for the effects of that inhibition. Because rats are more sensitive to the effects of inhibition of iodide uptake, the dose-response relationships observed in rat studies are not good estimates of the dose-response relationships in humans. The committee considered several of the animal studies on which EPA based its point of departure to be flawed in their design and execution. Conclusions based on those studies, particularly the neurodevelopmental studies, were not supported by the results of the studies (see Chapter 4 for a discussion of the animal studies). The committee also does not think that changes in brain morphometry, thyroid histopathology, and serum thyroid hormone concentrations should be used as the point of departure for the perchlorate risk assessment. Rather, the committee recommends that inhibition of iodide uptake by the thyroid, which is the key biochemical event and not an adverse effect, should be used as the basis of the risk assessment. Inhibition of iodide uptake is a more reliable and valid measure, it has been unequivocally demonstrated in humans exposed to perchlorate, and it is the key event that precedes all thyroid-mediated effects of perchlorate exposure.

The committee emphasizes that its recommendations differ from the traditional approach to deriving an RfD. The committee is recommending using a nonadverse effect rather than an adverse effect as the point of departure for the perchlorate risk assessment. Using a nonadverse effect that precedes the adverse effects is a conservative, health-protective approach to the perchlorate risk assessment, and the committee's recommendations for uncertainty factors reflect the conservatism of the approach.

The committee reviewed the human and animal data and found that the human data provided a more reliable point of departure for the risk assessment than the animal data (see Chapters 2, 3, and 4). The committee recommends using clinical data collected in a controlled setting with the relevant route of exposure to derive the RfD. Although the data from epidemiologic studies of the general population provide some information on possible effects of perchlorate exposure, those studies are ecologic and inherently limited with respect to establishing causality and serving as a basis of quantitative risk assessment. Furthermore, those studies typically focused on changes in serum thyroid hormone and TSH concentrations or clinical manifestations of the changes, not on inhibition of iodide uptake by the thyroid. Therefore, the committee is not recommending using the available epidemiologic studies to derive the point of departure for the risk assessment.

The committee recommends using the data from Greer et al. (2002) for derivation of the RfD. As discussed in Chapter 2, Greer et al. (2002) administered perchlorate at 0.007-0.5 mg/kg per day for 14 days to groups of healthy men and women. Inhibition of radioiodide uptake by the thyroid was measured 1 day before perchlorate administration, on days 2 and 14 of perchlorate administration, and 15 days after cessation of perchlorate administration, except that uptake was not measured on day 2 in the lowest-dose group. Serum thyroid hormones and TSH were measured before, during, and after perchlorate administration. The investigators found that inhibition of 24-hour radioiodide uptake by the thyroid ranged from 1.8% in the lowest-dose group to 67.1% in the highest-dose group. The inhibition was not significantly different from baseline in the lowest-dose group (0.007 mg/kg per day) but was significantly different from baseline in all other dose groups (0.02, 0.1, and 0.5 mg/kg per day). As discussed in Chapter 2, the very small decrease (1.8%) in thyroid radioiodide uptake in the lowest dose group was well within the variation of repeated measurements in normal subjects. Serum thyroid hormone concentrations did not change significantly in any group. Serum TSH concentrations decreased slightly and transiently in the highest-dose group—a change in the direction opposite what would be expected had thyroid hormone secretion decreased. The study identified a NOEL for inhibition of iodide uptake by the thyroid at 0.007 mg/kg per day, which is consistent with findings of similar studies in humans (Lawrence et al. 2000, 2001; Braverman et al. 2004), as described in Chapter 2.

The committee notes that the NOEL identified by Greer et al. (2002) (0.007 mg/kg per day) is lower than all the LOAELs and almost all the NOAELs identified by EPA in studies using rats, the most sensitive species studied (see Figure 5-3). Human equivalent values based on the animal data were all above 0.007 mg/kg per day (EPA 2003).

As part of its deliberations on the point of departure, the committee reviewed the BMD analyses conducted by EPA (2003), the California Environmental Protection Agency (CalEPA 2004), and Crump and Goodman (2003) on the data from Greer et al. (2002). Overall, those analyses used different models, approaches, parameters, response levels, and input data, so comparison of the results of the analyses is difficult. Although the committee recognizes that BMD modeling can be an improvement over the use of the NOAEL or LOAEL as a point of departure, there appears to be no consensus on the criteria for choosing one BMD approach over another. Because no clear justifications were provided with the individual analyses of the Greer et al. (2002) data that allowed selection of one set of results over another, the committee concluded that using the NOEL (0.007 mg/kg per day) for iodide uptake inhibition from Greer et al. (2002) as the point of departure provides a reasonable and transparent approach to the perchlorate risk assessment. As noted above, the NOEL value from Greer et al. (2002) is consistent with other clinical studies that have investigated iodide

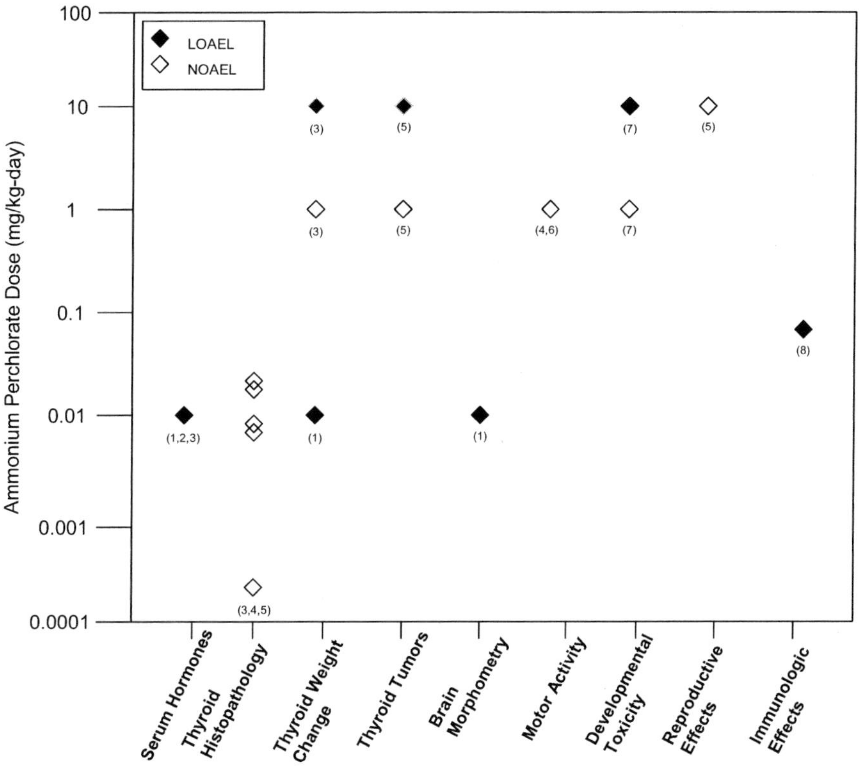

FIGURE 5-3 EPA's summary of NOAELs and LOAELs for various health effects in *rat* studies. Numbers indicate the following sources: 1, Argus Research Laboratories (2001), Consultants in Veterinary Pathology (2003); 2, Springborn Research Laboratories (1998); 3, Springborn Research Laboratories (1998); 4, Argus Research Laboratories (1998); 5, Argus Research Laboratories (1999); 6, Bekkedal et al. (2000); 7, Argus Research Laboratories (2000); 8, BRT-Burleson Research Technologies (2000a,b,c). Adapted from EPA 2003, p. 7-30.

uptake inhibition by perchlorate (Lawrence et al. 2000, 2001; Braverman et al. 2004). That the NOEL value from Greer et al. (2002) is a health-protective and conservative point of departure is supported by the results of a 4-week study of higher doses in normal subjects (Brabant et al. 1992; see Chapter 2) and extensive human and animal data that demonstrate that there will be no progression to adverse effects if no inhibition of iodide uptake occurs (see Figure 5-2). As discussed in Chapter 2, a sustained exposure at more than 0.4 mg/kg per day would most likely be required to cause a sufficient decline in iodide uptake and thyroid hormone production to result in adverse health effects in normal adults. That estimate is based on clinical studies and studies of long-term treatment of patients who had hyperthyroidism. Finally, the occupational and environmental studies described in Chapter 3 do not provide any evidence that would raise concerns about using the NOEL from Greer et al. (2002) as the point of departure for the perchlorate risk assessment.

UNCERTAINTY FACTORS

Five uncertainty factors are typically considered in calculating an RfD. Those factors account for interspecies differences, intraspecies differences, failure to establish a NOAEL, lack of chronic data, and other database gaps. In its draft risk assessment, EPA proposed a composite uncertainty factor of 300 to apply to its point of departure of 0.01 mg/kg per day, which was based on changes reported in brain morphometry, thyroid histopathology, and serum thyroid hormone concentrations in rats given perchlorate orally. Factors for intraspecies variability, use of a LOAEL, lack of chronic data, and other database gaps contributed to EPA's composite factor.

The committee cannot comment on the uncertainty factors that EPA selected for its primary analysis based on animal data, because the factors are related to the point of departure, and the committee is recommending a point of departure based on human data. As an ancillary analysis, EPA did derive an RfD based on data from the Greer et al. (2002) study in its draft risk assessment (EPA 2002a). In the following subsections, the committee provides its recommendations regarding the five uncertainty factors on the basis of its recommended point of departure (the NOEL from Greer et al. [2002] for inhibition of iodide uptake by the thyroid) and provides comments on EPA's selection of uncertainty factors used in its ancillary analysis. The committee recognizes that EPA did derive an RfD based on BMD modeling of the Greer et al. (2002) data in its comment-response document (EPA 2003), but using that approach would affect the selection of uncertainty factors. Therefore, the most appropriate comparisons are with the original analysis presented in EPA's draft risk assessment (EPA 2002a).

Interspecies Factor

When animal data are used as the basis of the point of departure, an adjustment is typically made for the possibility that humans are more sensitive than the selected test species. In the absence of data on the relative sensitivity of humans and animals, a default uncertainty factor of 10 is applied to the point of departure. The factor is often adjusted if data are available.

Because the committee's point of departure is based on the human data presented in Greer et al. (2002), no adjustment for interspecies extrapolation is needed. Therefore, the interspecies uncertainty factor should be 1.

Intraspecies Factor

There can be variability in responses among humans. The intraspecies uncertainty factor accounts for that variability and is intended to protect populations more sensitive than the population tested. In the absence of data on the range of sensitivity among humans, a default uncertainty factor of 10 is typically applied. The factor could be set at 1 if data indicate that sensitive populations do not vary substantially from those tested.

For the perchlorate risk assessment, potentially the most sensitive population is fetuses, particularly those of pregnant women who have hypothyroidism or iodide deficiency. In pregnant women who have undiagnosed hypothyroidism, perchlorate exposure could exacerbate the hypothyroidism by inhibiting iodide uptake by the thyroid. The National Health and Nutrition Examination Survey (NHANES III, 1991-1994; NCHS 1996) found that 4.6% of the U.S. population 12 years old and older had hypothyroidism (0.3% overt hypothyroidism and 4.3% subclinical hypothyroidism), in most cases not previously known (Hollowell et al. 2002). Similarly, in pregnant women who have iodide deficiency, the deficiency could be exacerbated by perchlorate exposure. In the NHANES III cohort, daily iodide intake was less than 50 μg in 15% of women of childbearing age and 7% of pregnant women (Hollowell et al. 1998). However, serum thyroid hormone and TSH concentrations of those who had a daily iodide intake less than 50 μg were similar to those of people who had a higher daily iodide intake (Hollowell et al. 2002; Soldin et al. in press). Thus, the data indicate that iodide deficiency in the U.S. population is mild, if it exists, so perchlorate exposure most likely would not exacerbate it. Nonetheless, the risk assessment should protect pregnant women who might have hypothyroidism or iodide deficiency.

Because Greer et al. (2002) studied healthy men and women, an intraspecies uncertainty factor greater than 1 is appropriate to provide protection for sensitive populations. Although EPA recommended a reduction in the default uncertainty factor from 10 to 3 for intrahuman variability in its draft risk assessment, the committee recommends use of a full factor of 10 to protect the most sensitive population—the fetuses of pregnant women who might have hypothyroidism or iodide deficiency. The committee views its recommendation as conservative and health-protective, especially given that the point of departure is based on a nonadverse effect that precedes the adverse effect in the continuum of possible effects of perchlorate exposure (see Figure 5-2).

LOAEL-to-NOAEL Extrapolation Factor

The risk assessment is intended to estimate an exposure at which no adverse effect will occur in humans. Historically, the RfD has been extrapolated from a NOAEL in long-term animal studies and, more recently, derived from BMD modeling. However, many studies fail to establish a NOAEL, and BMD modeling is not always possible or appropriate for some datasets. Therefore, a LOAEL must be used in the analysis instead of a NOAEL. The LOAEL-to-NOAEL uncertainty factor is designed to reduce the higher LOAEL to the lower NOAEL and is often 3 or 10.

As discussed, the committee is recommending that a NOEL for a nonadverse effect (inhibition of iodide uptake by the thyroid) be used as the basis for the perchlorate risk assessment. That recommendation is considered to be a more conservative and health-protective approach for the perchlorate risk assessment than traditional risk assessments that use a point of departure based on an adverse effect. In its 2002 draft risk assessment (EPA 2002a), EPA indicated that the NOEL identified in the Greer et al. (2002) study was a minimal LOAEL and applied a factor of 3 in its ancillary analysis. The committee disagrees with EPA's analysis. Again, inhibition of iodide uptake by the thyroid is not an adverse effect, and the small degree of inhibition (1.8%) observed in the subjects given 0.007 mg/kg per day was not statistically different from the baseline value. Accordingly, the lowest dose (0.007 mg/kg per day) was recognized as a NOEL by the committee. Therefore, the LOAEL-to-NOAEL uncertainty factor should be set at 1.

Subchronic-to-Chronic Extrapolation Factor

The RfD is intended to protect individuals from lifetime exposures. Therefore, the data must address the potential of long-term exposure to cause adverse effects. If only data on short-term exposures are available, a subchronic-to-chronic uncertainty factor of 10 is often used in the derivation of an RfD. Long-term animal toxicology studies are often the basis of the risk assessment, in which case an uncertainty factor of 1 is appropriate.

The committee recommends that the NOEL for inhibition of iodide uptake by the thyroid from a human study that involved a 14-day administration of perchlorate be used as the point of departure for the risk assessment. If that nonadverse biochemical event (see Figure 5-2) is used to derive the RfD, chronic exposure will have no greater effect than that resulting from short-term exposure. In fact, it may well have less effect because of the capacity of the pituitary-thyroid system to compensate for iodide deficiency by increasing iodide uptake. If some inhibition of iodide uptake by the thyroid did occur at the minimal dose proposed for the point of departure, data from humans indicate that longer exposures are not likely to result in a greater or more severe response. According to preliminary data presented in an abstract, administration of perchlorate to normal subjects for 6 months at 0.007 and 0.04 mg/kg per day had no effect on thyroid function (Braverman et al. 2004). Furthermore, occupational data suggest that long-term exposure of workers to perchlorate at up to 0.5 mg/kg per day does not have adverse effects on thyroid function (see Chapter 3). Perchlorate also does not accumulate in the body but is rapidly cleared via excretion in the urine even during chronic exposure.

In its ancillary analysis (EPA 2002a), EPA recommended a factor of 3 for duration of exposure, given that there are no chronic studies in the database. As indicated above, the committee does not agree with EPA's rationale. First, if inhibition of iodide uptake by the thyroid is duration-dependent, the effect should decrease rather than increase with time, because compensation would increase the activity of the sodium-iodide symporter and therefore increase iodide transport into the thyroid. Second, concerns that the duration of Greer et al. (2002) is not sufficient to observe the effects of changes in thyroid function are not valid, because the point of departure is selected to prevent those changes. If inhibition of iodide uptake by the thyroid does not occur, there will be no changes in thyroid function in the short or long term. Therefore, the committee recommends that a subchronic-to-chronic uncertainty factor of 1 be used in the risk assessment, with the understanding that the committee's recommended point of departure (inhibition of iodide uptake by the thyroid) is also used.

Adequacy-of-Database Factor

The adequacy of the database is typically described in terms of a specific set of animal toxicology studies. For example, chronic toxicity, reproductive toxicity, developmental toxicity, and carcinogenicity studies are typically required to have the highest confidence in a database of studies. However, mode-of-action studies and relevant human studies can eliminate the need for various animal studies. If critical studies have not been conducted, an uncertainty factor is often used to account for this deficiency.

In its ancillary analysis, EPA recommended a factor of 3 for database deficiencies. The committee does not agree and concludes that the database is adequate for deriving an RfD on the basis of inhibition of iodide uptake by the thyroid in humans. First, the database contains both human and animal data on that end point. Second, there is no evidence that perchlorate exposure has effects that do not result from inhibition of iodide uptake by the thyroid. Toxicology studies designed to identify the most sensitive effect of perchlorate exposure have indicated that the thyroid is the primary target of perchlorate exposure in rats, which is the most sensitive species studied (see Chapter 4). There is also no evidence that perchlorate administration to animals or humans causes systemic effects that are not mediated by the thyroid, with the exception of the toxic effects of very high doses given to patients with hyperthyroidism many years ago. Those adverse effects have not been described in any of the more recent studies in which lower doses of perchlorate were given to patients with hyperthyroidism for as long as 2 years. Finally, the absence of high-quality animal studies on outcomes that are downstream of iodide uptake inhibition, such as neurodevelopmental studies, is not relevant. Selection of the point of departure is designed to prevent the first step in the mode-of-action continuum, and studies on downstream events are not necessary. Therefore, the committee recommends that the database-adequacy uncertainty factor be set at 1, with the understanding that the committee's recommended point of departure is used to derive the RfD.[1]

---

[1]One committee member thought that the factor for database uncertainty should be greater than 1 and provided the following rationale:

> *The RfD is derived from a study in which a group of only seven healthy adults was given 0.007 mg/kg of perchlorate daily for 14 days (Greer et al. 2002). Although two other studies had similar results, the total number of subjects is still small. In addition to the small number of subjects, no chronic exposure studies have been published. An uncertainty factor of 3 could account for the uncertainty surrounding the small number of subjects and the absence of a long-term study.*

(Continued on next page)

SUMMARY

The committee concludes that EPA's mode-of-action model of perchlorate toxicity does not provide an accurate representation of events that follow changes in thyroid hormone and TSH production. The committee finds that a more realistic representation of effects of changes in serum thyroid hormone and TSH concentrations would be hypertrophy or hyperplasia of the thyroid, which might lead eventually to hypothyroidism. If perchlorate exposure did result in hypothyroidism, possible outcomes would be metabolic sequelae at any age and abnormal growth and development in fetuses or children. The committee notes that effects downstream of inhibition of iodide uptake by the thyroid have not been clearly demonstrated in any human population exposed to perchlorate, even at doses as high as 0.5 mg/kg per day.

The committee also differs with EPA regarding the definition of adverse effects associated with perchlorate exposure. The committee does not think that transient changes in serum thyroid hormone or TSH concentrations are necessarily adverse effects. The committee concludes that the first adverse health effect that could result from perchlorate exposure in the proposed continuum of effects would be hypothyroidism. However, hypothyroidism should not be used as the basis of the risk assessment.

The committee recommends that inhibition of iodide uptake by the

---

The other committee members provided the following response:

Although the committee acknowledges that the low-dose group (0.007 mg/kg per day) in Greer et al. (2002) had only seven subjects, the study examined the effects of four doses in a total of 37 subjects. In addition to the Greer et al. (2002) study, there are four other studies in which healthy adults were given perchlorate. The results of all the studies are remarkably similar (see Chapter 2, pp. 65-67). In addition to those studies, the studies of long-term treatment of hyperthyroidism and the studies of occupational and environmental exposure add confidence to the overall database. The issue concerning the absence of a long-term study is discussed in the section Subchronic-to-Chronic Extrapolation Factor in Chapter 5. Briefly, the key is recognizing that the committee is recommending that the RfD be based on inhibition of iodide uptake by the thyroid, a non-adverse biochemical event that precedes any adverse effects in the mode-of-action model. If that effect is used to derive the RfD, chronic exposure will have no greater effect than that resulting from short-term exposure, and in fact, it may well have less effect because of the capacity of the pituitary-thyroid system to compensate for iodide deficiency by increasing iodide uptake (see Chapter 5, p. 175).

thyroid, a nonadverse effect, be used for the point of departure in the perchlorate risk assessment. Using that biochemical event provides a conservative, health-protective approach to the risk assessment. If that event does not occur, all other proposed effects of perchlorate exposure would be avoided. Greer et al. (2002) provides a human dataset that can be used to derive the RfD, which is consistent with similar clinical studies. The committee recommends that the NOEL for inhibition of iodide uptake in that study, 0.007 mg/kg per day, be used as the point of departure for the risk assessment.

If the committee's recommendation is used as the point of departure, it recommends using a total uncertainty factor of 10. A full factor of 10 should be used for the intraspecies factor to protect the most sensitive population—the fetuses of pregnant women who might have hypothyroidism or iodide deficiency. No additional factors are needed for duration or database uncertainties. The database is sufficient, given the point of departure selected—one based on inhibition of iodide uptake by the thyroid.

The committee recognizes that its recommendations would lead to an RfD of 0.0007 mg/kg per day.[2] That value is supported by other clinical studies, occupational and environmental epidemiologic studies, and studies of long-term perchlorate administration to patients with hyperthyroidism. The committee concludes that an RfD of 0.0007 mg/kg per day should protect the health of even the most sensitive populations. The committee acknowledges that the RfD may need to be adjusted upward or downward on the basis of future research, such as that suggested in this report (see Chapter 6).

REFERENCES

Argus Research Laboratories, Inc. 1998. A Neurobehavioral Developmental Study of Ammonium Perchlorate Administered Orally in Drinking Water to Rats. ARGUS 1613-002. Argus Research Laboratories, Inc., Horsham, PA. [See also York, R.G., J. Barnett, W.R. Brown, R.H. Garman, D.R. Mattie, and D. Dodd. 2004. A rat neurodevelopmental evaluation of offspring, including evaluation of adult and neonatal thyroid, from mothers treated with ammonium perchlorate in drinking water. Int. J. Toxicol. 23(3):191-214.]

Argus Research Laboratories, Inc. 1999. Oral (Drinking Water) Two-Generation (One Litter Per Generation) Reproduction Study of Ammonium Perchlorate in

[2]For comparison, EPA's draft RfD in its 2002 draft risk assessment was 0.00003 mg/kg per day.

Rats. ARGUS 1416-001. Argus Research Laboratories, Inc., Horsham, PA.

Argus Research Laboratories, Inc. 2000. Oral (Drinking Water) Developmental Toxicity Study of Ammonium Perchlorate in Rats. ARGUS 1416-003D. Argus Research Laboratories, Inc., Horsham, PA.

Argus Research Laboratories, Inc. 2001. Hormone, Thyroid and Neurohistological Effects of Oral (Drinking Water) Exposure to Ammonium Perchlorate in Pregnant and Lactating Rats and in Fetuses and Nursing Pups Exposed to Ammonium Perchlorate During Gestation or Via Maternal Milk. ARGUS 1416-003. Argus Research Laboratories, Inc., Horsham, PA.

Bekkedal, M.Y.V., T. Carpenter, J. Smith, C. Ademujohn, D. Maken, and D.R. Mattie. 2000. A Neurodevelopmental Study of Oral Ammonium Perchlorate Exposure on the Motor Activity of Pre-Weanling Rat Pups. Report No. TOXDET-00-03. Neurobehavioral Effects Laboratory, Naval Health Research Center Detachment (Toxicology), Wright-Patterson Air Force Base, OH.

Brabant, G., P. Bergmann, C.M. Kirsch, J. Kohrle, R.D. Hesch, and von zur Muhlen. 1992. Early adaptation of thyrotropin and thyroglobulin secretion to experimentally decreased iodide supply in man. Metabolism 41(10):1093-1096.

Braverman, L.E., X. He, S. Pino, B. Magnani, and A. Firek. 2004. The effect of low dose perchlorate on thyroid function in normal volunteers [abstract]. Thyroid 14(9):691.

BRT-Burleson Research Technologies, Inc. 2000a. Ammonium Perchlorate: Effect on Immune Function. Quality Assurance Audit: Study No. BRT 19990524—Plaque-Forming Cell (PFC) Assay; Study No. BRT 19990525—Local Lymph Node Assay (LLNA) in Mice. BRT-Burleson Research Technologies, Inc., Raleigh, NC. June 30, 2000.

BRT-Burleson Research Technologies, Inc. 2000b. Addendum to Study Report: Ammonium Perchlorate: Effect on Immune Function. BRT 19990524 Study Protocol Plaque-Forming Cell (PFC) Assay; BRT 19990525 Study Protocol Local Lymph Node Assay (LLNA) in Mice. BRT-Burleson Research Technologies, Inc., Raleigh, NC. August 31, 2000

BRT-Burleson Research Technologies, Inc. 2000c. Ammonium Perchlorate: Effect on Immune Function. Study Report. BRT 19990524 Study Protocol Plaque-Forming Cell (PFC) Assay; BRT 19990525 Study Protocol Local Lymph Node Assay (LLNA) in Mice. BRT-Burleson Research Technologies, Inc., Raleigh, NC.

CalEPA (California Environmental Protection Agency). 2004. Public Health Goal for Chemicals in Drinking Water, Perchlorate. Office of Environmental Health Hazard Assessment, California Environmental Protection Agency. [Online]. Available: http://www.oehha.ca.gov/water/phg/pdf/finalperchlorate31204.pdf [accessed August 25, 2004].

Consultants in Veterinary Pathology, Inc. 2003. Hormone, Thyroid and Neurohistological Effects of Oral (Drinking Water) Exposure to Ammonium Perchlorate in Pregnant and Lactating Rats and in Fetuses and Nursing Pups Exposed to Ammonium Perchlorate During Gestation or Via Maternal Milk. Task One—

Review of Selective Morphometric Data From F1 Generation Day 22 Postpartum Rats, Including New Morphometric Data Obtained From Additional Step Sections. Morphometry Review Report. Protocol 1416-003. Consultants in Veterinary Pathology, Inc., Murrysville, PA. February 3, 2003.

Crump, K., and G. Goodman. 2003. Benchmark Analysis for the Perchlorate Inhibition of Thyroidal Radioiodine Uptake Utilizing a model for the Observed Dependence of Uptake and Inhibition on Iodine Excretion. Prepared for J. Gibbs, Kerr-McGee Corporation. January 24, 2003. (Presentation at the Fifth Meeting on Assess the Health Implications of Perchlorate Ingestion, July 29-30, 2004, Washington, DC.)

EPA (U.S. Environmental Protection Agency). 2000. Benchmark Dose Technical Guidance Document, External Review Draft. EPA/630/R-00/001. Risk Assessment Forum, U.S. Environmental Protection Agency, Washington, DC. [Online]. Available: http://www.epa.gov/ncea/pdfs/bmds/BMD-External_10_13_2000.pdf [accessed Nov. 17, 2004].

EPA (U.S. Environmental Protection Agency). 2002a. Perchlorate Environmental Contamination: Toxicological Review and Risk Characterization. External Review Draft. NCEA-1-0503. National Center for Environmental Assessment, Office of Research and Development, U.S. Environmental Protection Agency, Washington, DC. [Online]. Available: http://cfpub2.epa.gov/ncea/cfm/recordisplay.cfm?deid=23292 [accessed Aug. 25, 2004].

EPA (U.S. Environmental Protection Agency). 2002b. A Review of the Reference Dose and Reference Concentration Processes. Final Report. EPA/630/P-02/002F. Risk Assessment Forum, U.S. Environmental Protection Agency, Washington, DC. [Online]. Available: http://oaspub.epa.gov/eims/eimscomm.getfile?p_download_id=36836 [accessed Nov. 17, 2004].

EPA (U.S. Environmental Protection Agency). 2003. Disposition of Comments and Recommendations for Revisions to "Perchlorate Environmental Contamination: Toxicological Review and Risk Characterization, External Review Draft (January 16, 2002). [Online]. Available: http://cfpub2.epa.gov/ncea/cfm/recordisplay.cfm?deid=72117 [accessed Aug. 25, 2004].

EPA (U.S. Environmental Protection Agency). 2004. An Examination of EPA Risk Assessment Principles and Practices. EPA/100/B-04/001. Office of the Science Advisor, U.S. Environmental Protection Agency, Washington, DC. [Online]. Available: http://www.epa.gov/OSA/ratf-final.pdf [accessed Nov. 17, 2004].

Greer, M.A., G. Goodman, R.C. Pleus, and S.E. Greer. 2002. Health effects assessment for environmental perchlorate contamination: The dose response for inhibition of thyroidal radioiodine uptake in humans. Environ. Health Perspect. 110(9):927-937.

Hollowell, J.G., N.W. Staehling, W.H. Hannon, D.W. Flanders, E.W. Gunter, G.F. Maberly, L.E. Braverman, S. Pino, D.T. Miller, P.L. Garbe, D.M. DeLozier, and R.J. Jackson. 1998. Iodine nutrition in the United States. Trends and public health implications: Iodine excretion data from National Health and

Nutrition Examination Surveys I and III (1971-1974 and 1988-1994). J. Clin. Endocrinol. Metab. 83(10):3401-3408.

Hollowell, J.G., N.W. Staehling, W.D. Flanders, W.H. Hannon, E.W. Gunter, C.A. Spencer, and L.E. Braverman. 2002. Serum TSH, T4, and thyroid antibodies in the United States population (1988 to 1994): National Health and Nutrition Examination Survey (NHANES III). J. Clin. Endocrinol. Metab. 87(2):489-499.

ITER (International Toxicity Estimates for Risk Database). 2004. NOEL. ITER Definitions, ITER Glossary. Toxicology Excellence for Risk Assessment and Concurrent Technologies Corporation [Online]. Available: http://iter.ctcnet.net/publicurl/glossary.htm [accessed Nov. 10, 2004].

Lawrence, J.E., S.H. Lamm, S. Pino, K. Richman, and L.E. Braverman. 2000. The effect of short-term low-dose perchlorate on various aspects of thyroid function. Thyroid 10(8):659-663.

Lawrence, J., S. Lamm, and L.E. Braverman. 2001. Low dose perchlorate (3 mg daily) and thyroid function. Thyroid 11(3):295.

NCHS (National Center for Health Statistics). 1996. Third National Health and Nutrition Examination Survey: 1991-1994. U.S. Department of Health and Human Services, Centers for Disease Control and Prevention, National Center for Health Statistics, Hyattsville, MD.

Soldin, O.P., R.E. Trachtenberg, and J.C. Pezzullo. In press. Do thyroxine and thyroid-stimulating hormone levels reflect urinary iodine concentrations? Ther. Drug Monit.

Springborn Laboratories, Inc. 1998. A 90-Day Drinking Water Toxicity Study in Rats with Ammonium Perchlorate: Amended Final Report. SLI Study No. 3455.1. Springborn Laboratories, Inc., Spencerville, OH.

6

Research Recommendations

THE committee was asked to suggest scientific research that could reduce the uncertainty in the understanding of human health effects associated with low-dose perchlorate ingestion, especially research that could clarify "safe" exposures of sensitive populations. Rough estimates of timeframe, costs, and potential to reduce uncertainty were also requested for the proposed research. As noted in Chapter 5, the committee found that the data on perchlorate's mechanism of action and effects in animals and humans are sufficient to derive a reference dose (RfD). However, research that could provide a more complete understanding of the range of effects of perchlorate, especially effects of chronic exposure and effects on sensitive populations, was identified. The committee recommends a series of interrelated clinical, chronic toxicity, mechanistic, and epidemiologic studies that have the potential to clarify "safe" perchlorate exposures. The general scope and timeframe of the proposed studies are described; however, meaningful cost estimates cannot be generated in this report, because of the many variables that influence such calculations.

OVERVIEW

Recommendations for additional research fall into several categories. The first is to determine the effects in humans of chronic low-dose exposure to perchlorate with a prospective controlled clinical trial. The proposed study directly addresses the question of the capacity of the human thyroid to compensate for inhibition of thyroid iodide intake during long-term administration of perchlorate. The committee appreciates the controversial aspects of conducting human studies with perchlorate and provides an alternative approach that uses nonhuman primates if long-term studies in humans are not possible.

The effects of perchlorate on sensitive populations (fetuses, infants, and pregnant women) raise a number of fundamental questions, including the role of the sodium-iodide symporter (NIS) in placental iodide transport, the sensitivity of the NIS in the lactating breast to perchlorate inhibition, the influence of iodide status on perchlorate inhibition in placenta and breast, and finally direct effects of perchlorate on fetal development. Several models and methods developed in the mechanistic studies could be used in human investigations. Epidemiologic studies that include analysis of existing data, use of monitoring data, and new studies to determine perchlorate effects in sensitive populations are also proposed. Finally, further studies of the public-health implications of iodide status of pregnant women are proposed.

CHRONIC EXPOSURE

The proposed clinical study is designed to provide information on the potential chronic effects of perchlorate exposure on thyroid function focusing on the capacity for and mechanisms of thyroid compensation. The short-term human study (Greer et al. 2002) recommended as the point of departure in Chapter 5 reported inhibition of thyroid iodide uptake at perchlorate doses that are consistent with a wide array of human, animal, and in vitro studies. Long-term studies of populations that have iodide insufficiency, post-thyroidectomy patients, and workers occupationally exposed to perchlorate have demonstrated the large capacity of the thyroid to compensate for reduced iodide intake or thyroid mass. However, a rigorously designed controlled clinical study of prolonged exposure to perchlorate would clearly provide more specific information on the compensatory response to perchlorate exposure in humans and strengthen confidence in the RfD.

The hypothesis of the proposed clinical study is that administration of perchlorate in doses of 0.04 mg/kg per day or 0.1 mg/kg per day will transiently decrease thyroid iodide uptake but will have no long-term effect on thyroid function in healthy subjects. Those doses are based on the following data. Greer et al. (2002) reported that administration of perchlorate in doses of 0.02 and 0.1 mg/kg per day for 14 days to 10 healthy subjects per group reduced 24-hr thyroid iodide uptake by 16.4% and 44.7%, respectively (see Table 2-1 in Chapter 2). Braverman et al. (2004) reported that administration of 0.04 mg/kg per day to four healthy subjects had no effect on thyroid iodide uptake, measured at baseline and at 3 and 6 months, or on serum thyroid hormone and TSH concentrations, measured monthly. The small size of the study group in the latter study makes the results somewhat problematic, but the doses used provide a reasonable basis for a more comprehensive study to determine the effects of the dose at 0.04 mg/kg per day and the effects of a larger perchlorate dose, which also would inhibit iodide uptake acutely but for which compensation would be expected in the longer term.

The detailed protocol for this study is shown in Table 6-1. Briefly, it should be a double-blind study involving 90 healthy adults (45 men and 45 women). Subjects whose urinary iodide excretion exceeded 500 μg daily, indicative of a high iodide intake, would be excluded because of the antithyroid effects of a high iodide intake, but those with a low intake would not be excluded unless their urinary iodide excretion was less than 50 μg daily. Dietary iodide intake, as measured by urinary iodide excretion, would be monitored during the study but not controlled, given the high cost of the controlling the subjects' diet. After study selection, the participants should be randomly assigned to receive placebo or perchlorate at 0.04 or 0.1 mg/kg per day in drinking water. Thyroid function and general well-being should be monitored throughout the study. Assuming a 50% variance in mean serum TSH and a 20% variance in thyroid volume by ultrasound, a study with 30 subjects in each perchlorate treatment group and 30 subjects in the placebo group has greater than 90% power to detect a 100% increase in serum TSH and greater than 90% power to detect a 50% increase in thyroid volume by ultrasound at a 5% level of significance (Chow and Liu 1998). Given the costs of clinical testing and screening, subject payment, sample storage, and data analysis, the committee predicts that the study would cost at least $1.5 million.

If chronic studies in humans are not possible, chronic studies in nonhuman primates could provide useful information. Initial studies in monkeys could include a dose-range finding study with perchlorate and a 1-year low-dose chronic toxicity and thyroid function study. The chronic study could be designed to determine the effects of perchlorate administered in drinking water on thyroid iodide uptake and on thyroid gland function. Doses should be selected to evaluate effects of low-dose ingestion of perchlorate. Protocols for the suggested monkey studies are shown in Tables 6-2 and 6-3. Studies in pregnant monkeys could also provide useful information on the effects of perchlorate on fetal and neonatal development.

If additional studies are conducted in laboratory animals, including nonhuman primates, the committee recommends that additional research in support of physiologically based pharmacokinetic (PBPK) model develop-

TABLE 6-1 Protocol for Controlled Clinical Study

General Aspects • Design: Double blind • Number of subjects: 90 healthy adults (45 men and 45 women)
Selection of Subjects • Only subjects with normal examination and laboratory results should be included[a] • Subjects with 24-hr urinary iodide excretion over 500 µg to be excluded • Only nonpregnant women (confirmed by pregnancy testing before study) should be included; effective contraception must be used during study period • Subjects with high serum antithyroid antibody titers should not be excluded
Treatment Groups • Subjects randomly assigned to receive 1 of 2 potassium perchlorate doses or placebo for 6 months • Group size: placebo (15 men and 15 women), 0.04 mg/kg per day (15 men and 15 women), and 0.1 mg/kg per day (15 men and 15 women); these doses are predicted to reduce short-term 24-hr thyroid iodide uptake by about 25% and 45%, respectively • Potassium perchlorate and placebo administered in drinking water
Clinical Testing • Physical examination and measurements of serum free T_4, T_3, TSH, antithyroid peroxidase antibodies, and thyroglobulin; thyroid ultrasound at baseline, 1 week, 2 weeks, monthly • 24-hr urine collection for iodide at baseline, and 1, 3, and 6 months • 24-hr urinary and serum perchlorate at baseline, monthly to ensure compliance • 24-hr thyroid ^{123}I uptake at baseline, 2 weeks, 3 months, 6 months • Serum free T_4, T_3, TSH, antithyroid antibodies, and thyroglobulin; thyroid ultrasound; urinary iodide; urinary and serum perchlorate 1 month after perchlorate discontinuation • Ultrasound studies done without examiner knowledge of treatment group
Safety Monitoring • Data safety monitoring board should monitor thyroid function and ultrasound data • Subjects should be withdrawn from study if serum TSH rises to over 10 mU/L or thyroid volume increases by 100% over baseline at any time

[a]Decision is based on physical examination before study with evaluation of complete blood cell count, routine chemistries, thyroid function, antithyroid peroxidase antibody titers, and pregnancy status.

TABLE 6-2 Three-Month Dose-Range Finding Toxicity Study in Cynomolgus Monkeys

Dose Groups • Two of each sex in each group at 0 (control), low, middle, upper middle, and high dose • Ammonium perchlorate administered daily in drinking water for 3 months
Observations • General observations for morbidity and mortality, body weight, food consumption
Thyroid Function Assessment • Serum T_3, T_4, and TSH should be measured before dosing, on day 1, at weeks 1, 2, 4, 6, 8, 10, 12 • Thyroid ^{123}I uptake should be measured at appropriate times.
Clinical Test Measures • Comprehensive hematology and clinical chemistry before dosing, periodically during study
Pharmacokinetic Measures • Appropriate pharmacokinetic measures
Necropsy and Histopathology • Full necropsy, macroscopic examination, histopathologic examination of many tissues, including the thyroid and pituitary glands

ment, refinement, and validation be considered. As discussed in Appendix E, if additional studies are conducted in rats to elucidate the mode of action of perchlorate or dose-response relationships in potentially sensitive life stages (pregnant dams, fetuses, or neonates), a number of issues identified as potential data gaps with existing PBPK models could be addressed by studies that

- Develop a more biologically based description of placental transfer of perchlorate and iodide in rats.
- Determine whether perchlorate is transported by thyroid NIS if analytic methods of sufficient sensitivity can be developed or radiolabeled perchlorate with high radiochemical purity can be synthesized.
- Modify the adult human model to include the physiology of pregnancy and lactation to incorporate data from the recommended human clinical studies (if they are conducted).

TABLE 6-3 One-Year Chronic Toxicity Study in Cynomolgus Monkeys

Dose Groups • Six of each sex in each group at 0 (control), low, middle, upper middle, and high dose; dose selection based on findings of 3-month dose-range finding study • Ammonium perchlorate administered daily in drinking water for 12 months
Observations • General observations for morbidity and mortality, body weight, food consumption
Thyroid Function Assessment • Serum T_3, T_4, and TSH should be measured before dosing, on day 1, at weeks 1, 2, 4, monthly thereafter through 12 months • Thyroid ^{123}I uptake should be measured at appropriate time points
Clinical Test Measures • Comprehensive hematology and clinical chemistry before dosing and periodically during study
Pharmacokinetic Measures • Appropriate pharmacokinetic measures
Necropsy and Histopathology • Full necropsy, macroscopic examination, histopathologic examination of many tissues, including the thyroid and pituitary glands

- Modify models to incorporate dietary iodide measurements from biomonitoring studies in pregnant or lactating women, such as the studies of Soldin et al. (2003) and Hollowell et al. (1998).

By conducting a sensitivity analysis of their PBPK models, Clewell et al. (2003a,b), Merrill (2001), and Merrill et al. (2003) also identified several measures or biochemical processes that affect simulations of perchlorate and iodide disposition and target-tissue dosimetry in sensitive populations, including

- Clearance of perchlorate and iodide in urine.
- Rates of uptake and clearance of perchlorate and iodide in other tissues that contain NIS, such as skin, placenta, and mammary tissue.
- Protein binding of perchlorate in rat and human plasma at different life stages (adult, pregnant, neonate, and fetus).

- Time lag or kinetics for up-regulation of thyroid NIS by perchlorate (not just steady-state concentrations).
- Rates of production and secretion of thyroid hormones as related to iodide uptake by the thyroid in late-term fetal rats.

If those measures were assessed experimentally, the number of model parameters that need be estimated from animal studies could be reduced, and more data would then be available for model validation. Consideration should therefore be given to independently measuring some or all of the quantities, depending upon the types of simulations that are to be conducted in the future. Many of the studies, such as one to determine protein binding or others to improve models by incorporating data from the literature, can be conducted for under $50,000 and take only a few weeks or months to complete. Others—such as those to evaluate renal clearance, organification of iodide (thyroid hormone production and secretion) in the fetus, or placental transfer of perchlorate—could cost considerably more, perhaps $50,000-250,000, and involve several months or even a year or more to complete, depending on the study design.

If future studies are conducted in nonhuman primates, it will be important to develop a quantitative understanding of the disposition of perchlorate and iodide in the monkey model because cross-species extrapolations would still be required. In that case, the rat and human PBPK models would provide an initial template for development of a monkey model that would be supplemented by monkey-specific physiologic constants and biochemical constants similar to those already developed for rats and humans and identified through sensitivity analyses.

SENSITIVE POPULATIONS

Basic Studies

The identified sensitive populations are fetuses, infants, and pregnant women. Placental iodide transport and iodide concentration in the lactating breast have been widely studied with animal and in vitro models. Most of the studies, however, have not characterized in detail the specific contribution of the NIS or the influence of perchlorate inhibition. Although the NIS is expressed in the placenta, its contribution to iodide transfer from mother to fetus is not known. An especially critical issue in risk assessment has been the influence of iodide status on the sensitivity of placental and breast iodide transport to perchlorate.

Other tissues contain the NIS, such as the salivary glands, the gastric mucosa, and perhaps the choroid plexus. However, thyroid hormones are not produced in those tissues, and they are not essential even indirectly for thyroid hormone production, for example by their ability to transport iodide. The committee concludes that the highest priority should be given to studies to determine NIS function in the placenta and mammary gland because it is through these two tissues that iodide and possibly perchlorate reaches the thyroid gland in fetuses and infants.

The proposed studies suggested here use in vitro and animal models to determine the role of the NIS in placental iodide transport, the susceptibility of breast NIS to perchlorate inhibition, the role of iodide status on these effects, and the effects of perchlorate on development independent of effects on iodide transport.

Role of Placental NIS in Fetal Iodide Supply

Iodide availability to the developing fetus is likely to be influenced by a variety of factors, including maternal nutritional status, placental and uterine 5-deiodinase activity, and perhaps placental NIS activity. NIS mRNA and protein expression have been demonstrated in placenta and placenta-derived cells, but the specific role of placental NIS in iodide nutrition of the developing fetus is not known. NIS expression in the thyroid and lactating breast concentrates iodide to a gradient of 20:1 to 30:1. In the case of placental iodide transport, the usual condition is iodide moving down a concentration gradient from mother to fetus. Studies of human placenta and several animal models are proposed to study the role of the NIS in fetal iodide supply.

Normal human placenta expresses NIS mRNA and protein, but this has not been correlated with maternal iodide or thyroid status. Initial studies should characterize NIS mRNA and protein expression in the placenta, including cellular and subcellular localization, to provide the best correlation with iodide transport. A significant focus should be on functional measurement of iodide transport in placental tissue. Although a direct measurement of iodide transport in placenta would be most useful, it will be technically challenging because of the nature of the placenta. Placental tissue is a mixture of several cell types, which may express NIS differently. However, techniques are available to isolate specific cell types and could be used to study NIS activity in relevant cells more specifically. A functional assay of iodide transport in placental samples would be valuable for the later studies proposed to determine the impact of environmental exposure of pregnant women to perchlorate.

There are at least two important issues to address with respect to placental NIS in animal models: the physiologic role of placental NIS in supplying iodide to the developing fetus and the influence of perchlorate on this activity. A series of studies could be performed in mice; the potential to modify placental NIS gene expression make this model very useful. The studies should vary maternal dietary iodide intake and determine the influence on placental NIS mRNA and protein expression. The findings should be correlated with serum iodide concentration. The influence of perchlorate exposure could then be used to determine the influence on placental NIS expression and iodide transport. The studies should include measurement of maternal and fetal serum thyroid hormone and TSH. Litter size and neonatal development should be assessed. Expression of thyroid-hormone-dependent genes in the developing brain and behavior should be assessed.

Mammary Gland NIS and Perchlorate

Before identification and cloning of the NIS, an extensive literature characterized the factors that regulate the concentration of iodide in lactating breast. Lactogenic hormones, such as progesterone and prolactin, stimulate iodide uptake in lactating breast but not in nonlactating breast. Since the NIS has been cloned, the same factors have been shown to regulate NIS gene expression in the breast. The regulation of NIS expression in breast is different from that in the thyroid gland, which is regulated primarily by TSH. Although the same NIS protein is expressed in thyroid and breast, there have not been functional studies to determine the sensitivity of breast NIS to perchlorate inhibition.

The studies proposed should determine the influence of dietary iodide intake on NIS mRNA and protein expression and on cellular and subcellular localization of NIS in lactating mammary tissue. The iodide content of milk in lactating mice should be measured to reflect functional iodide transport. Once the function is assessed across a range of dietary iodide intakes, the influence of perchlorate can be tested. The effects of perchlorate are potentially on iodide transport and on thyroid function in the pups. Thyroid function in nursing pups should be evaluated, and thyroid-hormone-dependent genes in the brain and liver of the nursing pups profiled.

Tissue-Specific NIS Gene Inactivation

A key experiment to test the relative importance of NIS expression in placenta and lactating breast is to use tissue-specific inactivation of NIS gene expression. Methods are available by which NIS can be deleted from individual tissues, such as breast or placental tissue, in mice. Female mice that have NIS gene deletion can be used to characterize placental or lactating breast iodide transport and effects on their offspring. It is important to characterize the mice across a range of dietary iodide intake because the defects may be most apparent at lower iodide intake.

Scope of Basic Studies

The proposed studies, divided among laboratories with the appropriate expertise, would take 3-5 years to complete. The proposed studies represent four to six separate projects, each of which would require a "standard" basic investigation grant of $200,000 per year in direct costs. Thus, study costs range from $2.4 million (four projects for 3 years) to $6 million (six projects for 5 years) in direct costs.

Epidemiologic Research

The primary sources of uncertainty in estimating an RfD for perchlorate in drinking water arise from the absence of data on possible effects of exposure among populations at greatest risk of the adverse effects of iodide deficiency (pregnant women, their fetuses, and newborns). New epidemiologic research should focus on assessing possible health effects of perchlorate exposure in populations that are most vulnerable to the adverse effects of iodide deficiency. Studies should use direct measures of perchlorate exposure in individuals and methods—such as case-control, cohort, or nested designs—which are more suitable for examining potentially causal associations.

Future epidemiologic research on possible health effects of exposure to perchlorate in drinking water can be organized into additional analyses of existing data, new studies of health effects in selected populations, and monitoring of the frequencies of specific conditions in communities affected by the continuing efforts to reduce perchlorate in drinking water. Additional studies of existing data would be relatively inexpensive and could be completed in a few months. Because of the need for follow-up to assess developmental outcomes, new epidemiologic studies would take at least about 5-8 years to complete and cost millions of dollars.

Analyses of Existing Data

Several ecologic studies have examined the relation between birth in a community whose public drinking water contains perchlorate and perturbations of thyroid function in the newborn period (Brechner et al. 2000; Crump et al. 2000; F.X. Li et al. 2000; Z. Li et al. 2000; Schwartz 2001; Kelsh et al. 2003; Buffler et al. 2004). No study has specifically investigated a possible association of prenatal exposure to perchlorate and thyroid hormone abnormalities among low-birthweight babies (under 2,500 g), although this group of infants is thought to be potentially more vulnerable to the effects of such exposure. Two studies specifically excluded low-birthweight newborns (F.X. Li et al. 2000; Z. Li et al. 2000). It would be useful, where data are available, to compare neonatal thyroid hormone and TSH concentrations in low-birthweight infants in communities that have perchlorate contamination with thyroid hormone and TSH concentrations in low-birthweight babies in nonexposed communities. The data are already available from ecologic studies and could be used as a first step in investigating the relation of perchlorate exposure to perturbations of thyroid hormone secretion in this presumably sensitive group of neonates. Although the relation of perchlorate exposure to thyroid hormone secretion in preterm babies is also of interest, data on gestational age are not always available in the studies. In addition, gestational age is typically measured with less accuracy than birthweight.

New Studies in Selected Populations

Future studies of the health effects of perchlorate exposure should use case-control, cohort, or nested designs, in which data are obtained on exposure and outcome in the same people. Individual measures of perchlorate exposure can be difficult to obtain, but asking questions about the use of bottled water and the amount of tapwater consumed and taking measurements from the homes and schools of individual study participants (or from the workplaces of adults) would provide more direct measures of individual exposure and also account for variations in individual exposures in a given community. Urinary perchlorate might be assessed as a more direct measure of exposure. Using such measures in studies with adequate statistical power and close attention to control for confounding variables would address a number of the limitations in the available data.

Many of the important questions related to health effects have been addressed only by a single study or have not been adequately examined because inappropriate or insensitive end points have been used. For example, additional studies are needed to examine the possible relation between exposures to perchlorate during critical periods of neurodevelopment and adverse outcomes that reflect more subtle alterations in cognitive and motor function than are captured by the diagnostic labels "autism" and "attention-deficit hyperactivity disorder." Along those lines, the study in Chile should be extended to incorporate more extensive neurodevelopmental assessments of the children beginning in infancy and continuing through school age (Téllez, R.T., P.M. Chacón, C.R. Abarca, B.C. Blount, C.B. Van, K.S. Crump, J.P. Gibbs, Sótero del Rio Hospital, unpublished material, 2004). Functional end points that should be assessed in a follow-up of the infants include auditory function, including measures of oto-acoustic emissions; visual attention; cognitive function, including tests for executive function and memory; and tests of motor function, particularly balance, coordination, and rapid finger movements. The study being done in Chile also should be expanded to increase the sample sizes of the birth cohorts in each city. Larger sample sizes will be required for statistical power to be appropriate for detecting possible differences among exposure groups on the developmental assessments. It is also important to control for confounding variables. Little information is available as to whether or not exposures to perchlorate at concentrations present in municipal drinking water are related to an increased incidence of maternal hypothyroidism during gestation. That question could also be addressed in the current study of pregnant women in Chile, and use of larger samples would improve the precision of the estimates. Similar prospective studies done in the United States would be less useful for detecting subtle neurodevelopmental effects because perchlorate exposures are lower and the range of exposures is considerably narrower than those in Chile.

The cohort design is inherently limited, however, for examining outcomes in infants born to mothers who have gestational hypothyroidism. Such mothers, once identified in the study, would be treated, so their fetuses would no longer be subjected to their state of inadequate thyroid hormone. Potential effects of early gestational exposure of the fetuses to maternal hypothyroidism could still be examined, but those occurring beyond the first trimester probably could not be. In addition, the frequencies of some of the outcomes of interest, such as congenital hypothyroidism and perturbations of thyroid function in the newborn, are relatively uncommon; thus, cohorts would have to be very large to yield enough newborns with altered thyroid function for follow-up and sufficient power to detect meaningful developmental differences between exposure groups. Control for confounding factors also would be important.

Another way to investigate whether or not in utero perchlorate exposure increases the risk of adverse outcomes in the newborn (such as perturbations of thyroid hormones or congenital hypothyroidism) and later neurodevelopmental deficits, especially in infants of mothers who have hypothyroidism or dietary iodide deficiency, would be to use a hybrid nested case-control prospective design within birth cohorts. That design allows a focus on sets of newborns potentially at greatest risk for developmental abnormalities as a result of perchlorate exposure and later iodide deficiency. It also provides the opportunity to examine outcomes that have already occurred among mothers who had hypothyroidism during gestation. Such studies should be done in geographic areas that have different concentrations of perchlorate in drinking water, at least some of which are relatively high, as was done in the investigations in Chile.

For the nested case-control study, newborns who have abnormal thyroid hormone screening values (case group 1), those born at low birthweight or preterm (case groups 2 and 3), and a random sample of their birth cohort who have normal thyroid screening values and normal birthweights (controls) could be identified as soon as possible after birth from among all births in areas of known potential exposure to perchlorate. Data collection for the case groups and the controls should include interviews with mothers to collect information on sociodemographic variables and personal behavior; obstetrical history; usual consumption of drinking water and sources of water during the index pregnancy and after birth for drinking, formula preparation, and cooking; residential history during the index pregnancy; dietary intake of iodide-containing foods during pregnancy; and other relevant variables. Other information should include mothers' prenatal records for the index pregnancy, with special emphasis on whether there had been a diagnosis or treatment for hypothyroidism or hyperthyroidism or assessment of urinary iodide, and labor and delivery information. Blood samples from mothers and newborns should be obtained for measurement of postnatal serum thyroid hormones, TSH, and perchlorate. Urinary iodide could be measured in the mothers. Samples of water from the homes of cases and controls should be collected for measurement of perchlorate.

All case and control infants would then be followed prospectively with assessments of thyroid function, physical growth, and neurodevelopmental measures at appropriate ages beginning in infancy. The same measures recommended for follow-up of the cohort of Chilean newborns should be used. Test scores, growth measures, and frequencies of specific abnormal outcomes should then be compared in cases and controls. For example, the proportion of 5-year-olds who have abnormal performance on a test of motor function could be compared between cases (defined either on the basis of newborn thyroid hormone concentrations or birthweight) and controls born in areas of differing perchlorate exposure. Assessments should be done for a possible interaction between magnitude of perchlorate exposure and maternal hypothyroidism or iodide deficiency. The design also allows for identification of neurodevelopmental abnormalities, if any, associated with prenatal exposure to perchlorate in infants who do not have abnormalities of thyroid function or growth identifiable at birth.

Monitoring Studies

Remediation of perchlorate contamination in the Las Vegas wash is already being done, and average concentrations in the effluent have dropped significantly (Croft 2003). If perchlorate in public drinking water were a contributor to thyroid disease, perturbations in thyroid function, or developmental delays in exposed populations, one would expect a decrease in those outcomes after reduction in or cessation of the exposure. Continued monitoring of thyroid hormone concentrations in newborns and of the prevalence of thyroid diseases in populations that have already been studied for those outcomes would allow comparisons of disease frequencies before and after the remediation efforts were instituted. Such monitoring would be relatively inexpensive because it is based on existing, routinely collected data. Studies that examine time trends have limitations similar to those of ecologic studies, but they can provide indirect evidence of whether a particular exposure is related to a specific disease outcome by determining whether changes in exposure magnitude are associated with parallel changes in outcome frequency.

PUBLIC-HEALTH IMPLICATIONS OF IODIDE STATUS

In its deliberations on the health effects of perchlorate in drinking water, the committee considered pregnant women and fetuses to be particularly sensitive populations. Although iodide deficiency is believed to be rare in the United States, it has been reported in pregnant women, as has subclinical hypothyroidism and overt hypothyroidism (Klein et al. 1991; Hollowell et al. 1998). The committee believes that further research is needed to quantify more precisely the extent of and risk factors for iodide deficiency, particularly in pregnant women. However, while studies are being conducted, the committee emphasizes the importance of ensuring that all pregnant women have adequate iodide intake and, as a first step, recommends that consideration be given to adding iodide to all prenatal vitamin.

REFERENCES

Braverman, L.E., X. He, S. Pino, B. Magnani, and A. Firek. 2004. The effect of low dose perchlorate on thyroid function in normal volunteers [abstract]. Thyroid 14(9):691.

Brechner, R.J., G.D. Parkhurst, W.O. Humble, M.B. Brown, and W.H. Herman. 2000. Ammonium perchlorate contamination of Colorado River drinking water is associated with abnormal thyroid function in newborns in Arizona. J. Occup. Environ. Med. 42(8):777-782.

Buffler, P.A., M.A. Kelsh, E.C. Lau, C.H. Edinboro, and J.C. Barnard. 2004. Epidemiologic Studies of Primary Congenital Hypothyroidism and Newborn Thyroid Function Among California Residents. Final Report. University of California, Berkeley, CA. April 2004.

Chow, S.C., and J.P. Liu. 1998. Design and Analysis of Clinical Trials: Concepts and Methodologies. New York: Wiley and Sons.

Clewell, R.A., E.A. Merrill, K.O. Yu, D.A. Mahle, T.R. Sterner, D.R. Mattie, P.J. Robinson, J.W. Fisher, and J.M. Gearhart. 2003a. Predicting fetal perchlorate dose and inhibition of iodide kinetics during gestation: A physiologically based pharmacokinetic analysis of perchlorate and iodide kinetics in the rat. Toxicol. Sci. 73(2):235-255.

Clewell, R.A., E.A. Merrill, K.O. Yu, D.A. Mahle, T.R. Sterner, J.W. Fisher, and J.M. Gearhart. 2003b. Predicting neonatal perchlorate dose and inhibition of iodide uptake in the rat during lactation using physiologically based pharmacokinetic modeling. Toxicol. Sci. 74(2):416-436.

Croft, T. 2003. Overview of Las Vegas Valley Perchlorate Remedial Efforts. Presentation at the Second Meeting on Assess the Health Implications of Perchlorate Ingestion meeting, December 12 13, 2003, Irvine, CA.

Crump, C., P. Michaud, R. Tellez, C. Reyes, G. Gonzalez, E.L. Montgomery, K.S. Crump, G. Lobo, C. Becerra, and J.P. Gibbs. 2000. Does perchlorate in drinking water affect thyroid function in newborns or school-age children? J. Occup. Environ. Med. 42(6):603-612.

Greer, M.A., G. Goodman, R.C. Pleus, and S.E. Greer. 2002. Health effects assessment for environmental perchlorate contamination: The dose response for inhibition of thyroidal radioiodine uptake in humans. Environ. Health Perspect. 110(9):927-937.

Hollowell, J.G., N.W. Staehling, W.H. Hannon, D.W. Flanders, E.W. Gunter, G.F. Maberly, L.E. Braverman, S. Pino, D.T. Miller, P.L. Garbe, D.M. DeLozier, and R.J. Jackson. 1998. Iodine nutrition in the United States. Trends and public health implications: Iodine excretion data from National Health and Nutrition Examination Surveys I and III (1971-1974 and 1988-1994). J. Clin. Endocrinol. Metab. 83(10):3401-3408.

Kelsh, M.A., P.A. Buffler, J.J. Daaboul, G.W. Rutherford, E.C. Lau, J.C. Barnard, A.K. Exuzides, A.K. Madl, L.G. Palmer, and F. Lorey. 2003. Primary congenital hypothyroidism, newborn thyroid function, and environmental perchlorate exposure among residents of a southern California community. J. Occup. Environ. Med. 45(10):1116-1127.

Klein, R.Z., J.E. Haddow, J.D. Faix, R.S. Brown, R.J. Hermos, A. Pulkkinen, and M.L. Mitchell. 1991. Prevalence of thyroid deficiency in pregnant women. Clin. Endocrinol. 35(1):41-46.

Li, F.X., D.M. Byrd, G.M. Deyhle, D.E. Sesser, M.R. Skeels, S.R. Katkowsky, and S.H. Lamm. 2000. Neonatal thyroid-stimulating hormone level and perchlorate in drinking water. Teratology 62(6):429-431.

Li, Z., F.X. Li, D. Byrd, G.M. Deyhle, D.E. Sesser, M.R. Skeels, and S.H. Lamm. 2000. Neonatal thyroxine level and perchlorate in drinking water. J. Occup. Environ. Med. 42(2):200-205.

Merrill, E.A. 2001. Consultative Letter, AFRL-HE-WP-CL-2001-0008. PBPK Model for Perchlorate-Induced Inhibition of Radioiodide Uptake in Humans. Memorandum with attachments to Annie M. Jarabek, NCEA, U.S. Environmental Protection Agency, Research Triangle Park, NC, from Elaine Merrill, Air Force Research Laboratory/HEST, Department of the Air Force, Wright-Patterson Air Force Base, OH. June 5, 2001.

Merrill, E.A., R.A. Clewell, J.M. Gearhart, P.J. Robinson, T.R. Sterner, K.O. Yu, D.R. Mattie, and J.W. Fisher. 2003. PBPK predictions of perchlorate distribution and its effect on thyroid uptake of radioiodide in the male rat. Toxicol. Sci. 73(2):256-269.

Schwartz, J. 2001. Gestational Exposure to Perchlorate is Associated with Measures of Decreased Thyroid Function in a Population of California Neonates. M.S. Thesis, University of California, Berkeley.

Soldin, O.P., S.J. Soldin, and J.C. Pezzullo. 2003. Urinary iodine percentile ranges in the United States. Clin. Chim. Acta 328(1-2):185-190.

Appendix A

Biographic Information on the Committee to Assess the Health Implications of Perchlorate Ingestion

RICHARD B. JOHNSTON, JR. (*Chair*) is associate dean for research development and professor in the department of pediatrics at the University of Colorado School of Medicine and executive vice president for academic affairs at the National Jewish Medical and Research Center. His research interests include mechanisms of resistance to infection, the cell biology of neutrophils and macrophages, immune deficiency diseases, and child health. He has published more than 250 papers on these topics. Dr. Johnston has served as chair of pediatrics at the University of Pennsylvania, medical director of the March of Dimes, chief of pediatric immunology at Yale, member of the advisory committee for the National Center for Environmental Health of the Centers for Disease Control and Prevention, and president of three academic pediatric societies. He has previously chaired five Institute of Medicine (IOM) committees and has served on numerous other IOM committees and on the Board of Health Promotion and Disease Prevention. He was elected to IOM in 1995 and is a Fellow of the American Association for the Advancement of Science. He received his MD from Vanderbilt University.

STACY BRANCH is a consultant and owner of Djehuty Biomed Consulting and an adjunct associate professor with the Department of Animal Science in the School of Agriculture and Environmental Science at North Carolina Agricultural and Technical University. She was previously an associate professor of toxicology at North Carolina State University. Her research interests include molecular and classical teratology approaches to elucidate the mechanisms of xenobiotically induced abnormal mammalian development, including reproductive system development, and the nature of molecular pathways leading to abnormal phenotypes. Dr. Branch has published articles and book chapters on those subjects and forensic and clinical toxicology. She is managing editor of *Frontiers in Bioscience* and a member of the editorial board of the *Bulletin of Environmental Contamination and Toxicology*. Dr. Branch is a fellow of the American College of Forensic Examiners and a diplomate of the American Board of Forensic Medicine. She earned her DVM from Tuskegee University and her PhD in veterinary medical sciences (comparative biomedical sciences), pharmacology tract, from North Carolina State University.

GREGORY BRENT is professor of medicine and physiology at The David Geffen School of Medicine at the University of California, Los Angeles. Dr. Brent is also chief of the Endocrinology and Diabetes Section of the VA Greater Los Angeles Healthcare System and director of its fellowship program in endocrinology, metabolism, and diabetes. Dr. Brent's research interests include the molecular mechanisms of thyroid hormone action, regulation of iodide uptake in thyroid and breast cancer, the influence of thyroid hormone on development, and thyroid disease in pregnancy. He is currently secretary of the American Thyroid Association. Dr. Brent has been an associate editor of the journal *Thyroid*, and he remains on the editorial board. He recently completed a term on the National Institutes of Health Endocrinology Study Section and chaired a Special Endocrinology Study Section. Dr. Brent earned his MD from the University of Southern California.

ROSALIND BROWN is associate professor of pediatrics at Harvard University School of Medicine and director of clinical trials research in endocrinology at Children's Hospital in Boston. She was formerly professor of pediatrics at the University of Massachusetts Medical School and director of the Division of Pediatric Endocrinology and Diabetes at University of Massachusetts Memorial Health Center. Dr. Brown has published numerous articles, editorials, and book chapters with a special interest in thyroid development and disease and its consequences in childhood. She is a coeditor of the forthcoming 5th edition of *Clinical Endocrinology*. She has been a member of the editorial boards of the *Journal of Clinical Endocrinology and Metabolism* and *Thyroid*. Dr. Brown earned her MD from McGill University and is board-certified in pediatric endocrinology.

CHARLES C. CAPEN is distinguished university professor and former chairman (1981-2002) of the Department of Veterinary Pathobiology (restructured as Veterinary Biosciences in 1994) at the Ohio State University. His research interests include the effects of environmental pollutants on thyroid and ovarian function in rodents and secondary mechanisms of oncogenesis, the comparative aspects of endocrine and metabolic diseases, the gene transfer of the sodium iodide symporter in prostate and mammary cancer, and humoral factors in cancer-associated hypercalcemia. He has consulted in the past on thyroid issues for the U.S. Environmental Protection Agency, the Food and Drug Administration, other federal agencies, the International Agency for Research on Cancer, and private clients. Dr. Capen was elected to the Institute of Medicine in 1992 and is a diplomate and past president of the American College of Veterinary Pathologists. He earned his DVM from Washington State University and his PhD from the Ohio State University.

DAVID COOPER is professor of medicine-endocrinology at The Johns Hopkins University School of Medicine. He is also professor of international health at The Johns Hopkins University Bloomberg School of Hygiene and International Health, director of the Thyroid Clinic at The Johns Hopkins Hospital, and director of the Division of Endocrinology at the Sinai Hospital of Baltimore. His primary research interests are thyroid cancer, hyperthyroidism, and antithyroid drug pharmacology. Dr. Cooper earned his MD from Tufts University.

RICHARD CORLEY is staff scientist in the biomonitoring and biologic modeling group at the Pacific Northwest Laboratory operated by Battelle for the U.S. Department of Energy in Richland, Washington. Dr. Corley specializes in the development of physiologically based pharmacokinetic models, real-time breath analysis, dermal and inhalation bioavailability, and the development of three-dimensional computational fluid-dynamic models of the respiratory system. He has published numerous peer-reviewed papers on oral, dermal, and inhalation toxicology; modes of action of a variety of industrial and consumer chemicals; and pharmacokinetic modeling and its applications in human health risk assessments. He received his PhD in environmental toxicology from the University of Illinois at Urbana-Champaign.

LINDA COWAN is George Lynn Cross Research Professor in the Department of Biostatistics and Epidemiology at the University of Oklahoma Health Sciences Center. Her research interests include cardiovascular disease and the relative importance of different risk factors in men and women and American Indian populations, neurologic disorders, perinatal epidemiology, and the application of epidemiology in the legal setting. Her recent research includes evaluating risk factors for abnormal fetal growth and adverse neurologic outcomes in infants and children and the role of inflammatory mediators in the pathology of the nervous system from infancy through old age. Dr. Cowan has served on several committees of the Institute of Medicine (IOM) and is a member of the Advisory Board of the IOM Medical Follow-up Agency. She earned her PhD in epidemiology from The Johns Hopkins University.

JAMES LAMB IV is senior vice president at The Weinberg Group Inc. He was previously senior vice president at Blasland, Bouck & Lee Sciences, Inc. His interests include risk assessment, general toxicology, carcinogenesis, and reproductive and developmental toxicology. Dr. Lamb has published extensively in those fields and in pesticide regulation and pathology. He has served on two National Research Council committees: the Committee on Risk Characterization and the Committee on Hormonally Active Agents in the Environment. Dr. Lamb is a past president and diplomate of the American Board of Toxicology. He received his PhD in pathology from the University of North Carolina at Chapel Hill and a JD from the North Carolina Central University School of Law.

GEORGE LAMBERT is the director of the Center for Childhood Neurotoxicology and Exposure Assessment at the Environmental and Occupational Health Sciences Institute. The center is supported by competitively awarded grants from National Institute of Environmental Health Sciences and the Environmental Protection Agency, and the institute is sponsored by Rutgers, the State University of New Jersey, and the University of Medicine and Dentistry of New Jersey (UMDNJ)-Robert Wood Johnson Medical School. Dr. Lambert is also an associate professor of pediatrics and director of the Division of Pediatric Pharmacology and Toxicology at UMDNJ-Robert Wood Johnson Medical School and is an attending neonatologist. His current research is focused on the influences of environmental exposure to neurotoxicants on children's neurologic health, the presence of plasticizers in the human newborn, the effects of environmental endocrine disruption on hypospadism and cryptorchidism in children, in utero exposure to environmental chemicals and reproductive function in men, and the role of gene polymorphisms in birth defects. Dr. Lambert earned his MD from the University of Illinois at Chicago and completed his residency in pediatrics at The Johns Hopkins Hospital and a research fellowship in molecular teratology at the National Institute of Child Health and Human Development.

R. MICHAEL MCCLAIN is a consultant in toxicology with McClain Associates and adjunct professor in the Environmental and Occupational Health Sciences Institute at the University of Medicine and Dentistry of New Jersey. Previously, he was distinguished research leader with Hoffmann-LaRoche, Inc. Dr. McClain is a diplomate of the American Board of Toxicology, a fellow of the Academy of Toxicological Sciences, and past president of the Society for Toxicology. He is experienced in teratology, reproductive toxicology, general toxicology, and carcinogenicity testing. His research activities include the mechanisms of chemical carcinogenesis in the thyroid, liver, and adrenals and the regulatory aspects of cancer risk assessment. He reviewed scientific studies on perchlorate for private clients and provided comments to the Environmental Protection Agency Peer Review Panel on Perchlorate in February 1999. Dr. McClain received his PhD in pharmacology from the University of Iowa.

SUSAN SCHANTZ is professor of toxicology at the University of Illinois at Urbana-Champaign. She is director of a Children's Environmental Health Research Center and a National Institute of Environmental Health Sciences (NIEHS) Environmental Toxicology Training Program. The center is supported by competitively awarded grants from NIEHS and the Environmental Protection Agency. She also chairs the Interdisciplinary Environmental Toxicology Program at the university. Dr. Schantz's research interests include the effects of exposure to environmental neurotoxicants during development and aging—particularly the nervous system effects of polychlorinated biphenyls, dioxins, and related compounds—and modeling human exposures to identify which chemicals mediate neuropsychologic effects. Dr. Schantz earned her PhD in environmental toxicology from the University of Wisconsin, Madison.

DALENE STANGL is director of the Institute of Statistics and Decision Sciences and professor of the practice of statistics and public policy at Duke University. Her research interests include analyzing and promoting a stronger link between statistical analysis and decision-making and promoting Bayesian statistical methods in health-related research. She is a fellow of the American Statistical Association. Dr. Stangl has coedited two books: *Bayesian Biostatistics* (Marcel Dekker, 1996) and *Meta-Analysis in Medicine and Health Policy* (Marcel Dekker, 2000). She earned her PhD in statistics from Carnegie Mellon University.

LYNETTE STOKES is chief of the Bureau of Hazardous Materials and Toxic Substances in the Washington, DC, Environmental Health Administration within the Department of Health. Dr. Stokes has developed protocols and questionnaires for a cohort study of young adults to investigate neurologic, kidney, and reproductive outcomes associated with childhood lead exposure; designed and conducted epidemiologic analyses with univariate and multivariate methods; examined environmental factors associated with asthma; and copublished one of the first studies showing the relationship between childhood lead exposure and increased blood pressure in adults. She has also published on the neurotoxic effects of pesticides among agricultural workers. Previously, Dr. Stokes was an epidemiologist with the Centers for Disease Control and Prevention Agency for Toxic Substances and Disease Registry. She received her MPH in epidemiology from the University of Michigan, Ann Arbor, and PhD in environmental health and toxicology (concentration in epidemiology) with a minor in neurotoxicology from the State University of New York, Albany.

ROBERT D. UTIGER is clinical professor of medicine at the Harvard University School of Medicine and editor-in-chief of *Clinical Thyroidology*. He was a deputy editor of the *New England Journal of Medicine* from 1989 to 2000 and was editor-in-chief of the *Journal of Clinical Endocrinology and Metabolism* from 1983 to 1989. He is coeditor of *Werner and Ingbar's The Thyroid: A Fundamental and Clinical Text*. He previously served on the faculty of the Washington University School of Medicine, the University of Pennsylvania School of Medicine, and the University of North Carolina School of Medicine. Dr. Utiger has extensive expertise in thyroid disease and pituitary-thyroid physiology. He is a master of the American College of Physicians. He served on the Institute of Medicine Committee on Medicare Coverage of Routine Thyroid Screening. Dr. Utiger earned his MD from the Washington University School of Medicine.

Appendix B

Glossary

Argus: Argus Research Laboratories. A division of Charles River Discovery and Development Services, a contract research organization. Argus has 25 years of experience in the conduct of nonclinical reproductive and developmental toxicity safety studies for the pharmaceutical, food, and chemical industries. Argus has a specialization in neurobehavioral studies, a standard component of reproductive and developmental toxicity tests for pharmaceutical products and neurotoxicology safety assessments.

AUC: area under the curve. A measure of exposure that includes duration and concentration. It is calculated from the curve that results when the concentrations of the test substance in some biologic tissue, typically blood, are plotted against the exposure time.

BMD: benchmark dose. A dose or concentration that produces a selected change in the response rate or occurrence of an adverse effect or other relevant end point compared with background.[1]

BRT: Burleson Research Technologies, Inc. A for-profit contract laboratory specializing in proof-of-concept, preclinical, and toxicology studies.

BW: body weight.

caliche: a sedimentary deposit rich in sodium nitrate and other soluble salts.[2] Sodium perchlorate has been found in caliche from Chile.

CI: confidence interval. The Frequentist statistical confidence interval describes the relative frequency with which a parameter's values lie within a calculated interval in a series of repeat samples.[3]

ClO_4^-: perchlorate.

colloid: the stored secretion in follicles of the thyroid gland.[4]

credible interval: Bayesian credible interval describes an interval of the parameter space such that the probability that the parameter's value lies in the interval is at least the given percentage.

CV: coefficient of variation. A metric useful in comparing the variability of multiple datasets calculated by dividing the standard deviation of a dataset by its mean.[5]

deiodinases: enzymes that metabolize thyroid hormones, including conversion of the prohormone thyroxine to triiodothyronine by removal of an iodide.

diiodotyrosine: an intermediate in the production of thyroid hormones. It contains two iodine atoms.[4]

dL: deciliter. One-tenth of a liter or 100 milliliters.

DL: day of lactation.

ecologic study: a type of observational epidemiologic study in which the units of analysis are populations or groups of people rather than individuals.

ED_{10}: effective dose 10%. The ED_{10} is the dose associated with a 10% increase in an adverse effect compared with the control.[1]

GD: gestation day.

geometric mean: the nth root of the product of n observations.[5] The geometric mean is similar to the median and always less than the arithmetic mean unless the observations are identical, making the geometric mean equal to the arithmetic mean.[6]

goiter: a chronic enlargement of the thyroid gland, not due to a neoplasm.[4]

goitrogen: any substance that induces goiter.[4]

HEE: human equivalent exposure.

HPT: *see* hypothalamic-pituitary-thyroid axis.

hyperplasia: an increase in the number of normal cells in a tissue or organ, excluding tumor formation, whereby the bulk of the part or organ may be increased.[4]

hyperthyroidism: increased secretion of thyroxine and triiodothyronine from the thyroid gland.

hypertrophy: general increase in bulk of a part of organ, not due to tumor formation.[4]

Hypothalamic-pituitary-thyroid axis: the body's regulatory system for the thyroid. Thyrotropin-releasing hormone from the hypothalamus stimulates the pituitary gland to secrete thyroid-stimulating hormone, which acts on the thyroid gland to stimulate thyroid hormone synthesis and secretion. High concentrations of thyroid hormone in the blood feed back to the hypothalamus and pituitary and inhibit thyroid-stimulating hormone secretion.[4]

hypothyroidism: decreased secretion of thyroxine and triiodothyronine from the thyroid gland.

I^-: iodide. Iodide is the physiologically active anion of iodine. It is a micronutrient essential for thyroid hormone production.[7]

iodine: an atomic element that occurs nutritionally in the form of iodide.

IUI: iodide uptake inhibition.

K_m: *see* Michaelis-Menten constant.

L/hr per kg: liters per hour per kilogram of body weight.

LOAEL: lowest observed-adverse-effect level. The lowest exposure at which there are statistically or biologically significant increases in frequency or severity of adverse effects in an exposed population compared with its appropriate control group.[8]

mechanism or mode of action: processes causing a biologic effect, for example, toxicity.

mg: milligram (one-thousandth of a gram).

mg/hr/kg: milligrams per hour per kilogram of body weight.

mg/kg: milligrams per kilogram of body weight.

mg/L: milligrams per liter (equivalent to parts per million [ppm]).

Michaelis-Menten constant: the chemical concentration that can be saturated at half maximal capacity of a physiologic process, such as protein binding or active transport.[9]

milliunits per liter: a measure of concentration based on the biologic activity of a substance rather than its weight or volume.

mL: milliliter (one-thousandth of a liter).

mM: millimolar (concentration of 1 millimole per liter).

monoiodotyrosine: an intermediate in the production of thyroid hormones that contains a single iodine atom.[4]

mRNA: messenger ribonucleic acid.

mU/L: *see* milliunits per liter.

μg: microgram (one-millionth of a gram).

μg/dL: micrograms per deciliter.

μg/L: micrograms per liter (equivalent to parts per billion [ppb]).

μM: micromolar (a concentration of 1 micromole per liter).

ng: nanogram (one-billionth of a gram).

ng/hr/kg: nanograms per hour per kilogram of body weight.

ng/L: nanograms per liter.

NIS: *see* sodium (Na^+)/iodide(I^-) symporter.

NOAEL: no-observed-adverse-effect level. An exposure level at which there are no statistically or biologically significant increases in the frequency or severity of adverse effects in an exposed population compared with its appropriate control; some effects may be produced at this level, but they are not considered as adverse nor precursors to specific adverse effects. In an experiment with several NOAELs, the regulatory focus is primarily on the highest one; this leads to the common usage of the term NOAEL as the highest exposure without adverse effect.[8]

NOEL: no observed-effect level. An exposure level at which there are no statistically or biologically significant increases in the frequency or severity of any effect between the exposed population and its appropriate control.[1]

OR: odds ratio. A measure of association between exposure and disease development calculated by dividing the odds that an exposed group develops the disease by the odds that a nonexposed group develops the disease.[10]

PBPK model: *see* physiologically based pharmacokinetic.

PCB: polychlorinated biphenyl.

PD: pharmacodynamic.

physiologically based pharmacokinetic model: a type of model that estimates the dose delivered to a target tissue or organ by taking into account the rate of absorption into the body, distribution among target organs and tissues, metabolism, and excretion.[1]

PND: postnatal day.

point of departure: the dose-response point that marks the beginning of an extrapolation.[1]

posterior probability distribution: *See* prior probability distribution.

ppb: parts per billion.

ppm: parts per million.

prior probability distribution: In a Bayesian statistical approach, expert judgment and available data are modeled in a prior probability distribution. Bayes' theorem can be applied to update the prior probability distribution with data from new studies to yield a posterior probability distribution.

PTU: propylthiouracil.

PWG: Pathology Working Group.

RAIU: radioactive iodide uptake.

reference dose: an estimate (with uncertainty spanning perhaps an order of magnitude) of a daily oral dose delivered to the human population (including sensitive groups) that is likely to have no appreciable risk of deleterious effects during a lifetime. It can be derived from a NOAEL, LOAEL, or benchmark dose; uncertainty factors can be applied to reflect limitations of the data used. Generally used in EPA's noncancer health assessments.[1]

relative risk: a measure of association between an exposure and development of disease calculated by dividing the disease incidence in an exposed group by the disease incidence in a nonexposed group.[10]

RfD: *see* reference dose.

RR: *see* relative risk.

RSC: relative source contribution. An estimate of the percentage of a toxicant exposure coming from a particular source, such as drinking water or food. Determining the relative source contribution is an important part of risk assessment and risk management when the toxicant is found in more than one source. Because the RfD is a total daily dose, if the toxicant appears in both water and food, for example, the portion of the exposure dose from each source is determined to set appropriate food and water standards so as not to exceed the RfD.

SIR: *see* standardized incidence ratio.

SMR: *see* standardized mortality ratio.

sodium (Na^+)/iodide(I^-) symporter: a plasma membrane glycoprotein that mediates the active transport of iodide and is expressed in a variety of tissues, including thyroid, salivary glands, lactating breast, and placenta. A sodium gradient maintained by Na (sodium)-K (potassium) adenosine triphosphatase is required for iodide transport by NIS.[11]

SPR: *see* standardized prevalence ratio.

standardized incidence ratio: a measure of population health calculated by dividing the observed disease incidence in a population of interest by the disease incidence expected on the basis of disease rates of a reference population.

standardized mortality ratio: a measure of population health calculated by dividing the number of deaths observed in a population of interest by the number of deaths expected on the basis of the mortality rates of a reference population.[12]

standardized prevalence ratio: a measure of population health calculated by dividing the observed disease prevalence in a population of interest by the prevalence expected in the population on the basis of disease rates of a reference population.

T_3: *see* triiodothyronine.

T_4: tetraiodothyronine. *See* thyroxine.

TBG: *see* thyroxine-binding globulin.

TCE: trichloroethylene.

TERA: Toxicology Excellence for Risk Assessment, a nonprofit corporation dedicated to the best use of toxicity data for risk assessment. TERA's mission is to protect public health by developing and communicating risk-assessment values, by offering consulting services to government and industry, improving risk methods through research, and educating the public on risk assessment issues.[13]

tetraiodothyronine: *see* thyroxine.

thyroglobulin: a thyroid protein that serves as the substrate for production of thyroxine and triiodothyronine and as the storage form of thyroid hormones.

thyroid: a butterfly-shaped gland in the front of the neck. It is composed of spherical follicles that are the functional units of the gland.

thyroid follicle: the functional unit of the thyroid gland, composed of a single layer of thyroid cells surrounding a lumen that contains thyroglobulin, the storage form of thyroxine and triiodothyronine.

thyroid peroxidase: an enzyme that catalyzes iodide organification into thyroglobulin and then promotes thyroid hormone synthesis.

thyrotropin: a glycoprotein hormone that stimulates growth and function of the thyroid gland. Also known as thyroid-stimulating hormone.[4]

thyrotropin-releasing hormone: a hormone that is released from the hypothalamus and stimulates synthesis and secretion of thyroid-stimulating hormone (TSH).[14]

thyroxine: a biologically inactive prohormone containing four iodine atoms that is activated to triiodothyronine by deiodinase. Also known as tetraiodothyronine.

thyroxine-binding globulin: the primary serum protein that binds thyroid hormones in humans.

TPO: *see* thyroid peroxidase.

TRC Companies, Inc.: a for-profit consulting organization specializing in energy, environment, infrastructure, and infrastructure security projects.[15]

TRH: *see* thyrotropin-releasing hormone.

triiodothyronine: thyroid hormone containing three iodine atoms. About 80% of the triiodothyronine produced in the body each day is formed outside the thyroid gland by removal of one iodine atom from thyroxine. Triiodothyronine acts in different organs by binding to nuclear thyroid hormone receptors and stimulates or inhibits gene expression.

TSH: thyroid-stimulating hormone. *See* thyrotropin.

tyrosine: an amino acid required for the production of thyroid hormones.

UF: uncertainty factor. Factors—typically 1, 3, or 10—used to derive an RfD from experimental data. The factors are intended to account for the variation in sensitivity among the members of the human population (interindividual variability), the uncertainty in extrapolating animal data to the human population (interspecies uncertainty), the uncertainty in extrapolating from data obtained in a study involving less-than-lifetime exposure (extrapolating from subchronic to chronic exposure), the uncertainty in using LOAEL data rather than NOAEL data, and the uncertainty associated with extrapolation when the database is incomplete.[16] A UF of 10 is considered to be a health-protective default value to be used when little is known about a particular source of variability or uncertainty or when information on a relevant health effect is lacking. As additional research becomes available, UFs change as indicated by the new information.

VGI: volume of gastrointestinal tract.

V_{max}: describes, in units of milligrams per hour, the maximal capacity or velocity for binding or transport, such as binding or transport of iodide by the NIS.

$V_{max}C$: describes, in units of milligrams per hour per kilogram of body weight, the maximal capacity or velocity for binding or transport scaled by body weight according to the equation: V_{max} (mg/hr) = [$V_{max}C$ (mg/hr per kg)][body weight (kg)]$^{0.70}$.

VSk: volume of skin.

NOTES

1. EPA (U.S. Environmental Protection Agency). 2003. Glossary of IRIS Terms. Integrated Risk Information System (IRIS). [Online]. Available http://www epa gov/iris/gloss8.htm [accessed April 16, 2004].

2. CanSIS (Canadian Soil Information System). 1996. Glossary: Caliche. Canadian Soil Information System, Canada Department of Agriculture, Ottawa. [Online]. Available: http://sis.agr.gc.ca/cansis/glossary/index.html [accessed April 16, 2004].

3. Colton, T. 1974. Statistics in Medicine, 1st Ed. Boston: Little, Brown and Company.

4. Stedman, T.L. 2000. Stedman's Medical Dictionary, 27th Edition. Philadelphia: Lippincott Williams and Wilkins.

5. Rosner, B. 1995. Pp. 23-24 in Fundamentals of Biostatistics. Belmont, CA: Duxbury Press.

6. Altman, D.G. 1991. Practical Statistics for Medical Research. London: Chapman and Hall.

7. IOM (Institute of Medicine). 2000. Iodine. Pp 258-289 in Dietary Reference Intakes for Vitamin A, Vitamin K, Arsenic, Boron, Chromium, Copper, Iodine, Iron, Manganese, Molybdenum, Nickel, Silicon, Vanadium, and Zinc. Washington, DC: National Academy Press.

8. NRC (National Research Council). 2000. Toxicological Effects of Methylmercury. Washington, DC: National Academy Press.
9. Medinsky, M.A., and C.D. Klaassen. 1996. Toxicokinetics. Pp. 187-198 in Casarett and Doull's Toxicology: The Basic Science of Poisons, C.D. Klaassen, ed. New York: McGraw-Hill.
10. Gordis, L. 1996. Epidemiology. Philadelphia: W.B. Saunders.
11. Dohan, O., A. de la Veija, V. Paroder, C. Riedel, M. Artani, M. Reed, C.S. Ginter, and N. Carrasco. 2003. The Sodium/Iodide Symporter (NIS): Characterization, regulation, and medical significance. Endocr. Rev. 24(1): 48-77.
12. Mausner, J.S., S. Kramer, and A.K. Bahn. 1985. Epidemiology: An Introductory Text, 2nd Ed. Philadelphia: W.B. Saunders Company.
13. Toxicology Excellence for Risk Assessment (TERA). 2004. TERA Homepage. [Online]. Available: http://www.tera.org/ [accessed April 16, 2004].
14. Beers, M.H., and R. Berkow, eds. 1999. The Merck Manual of Diagnosis and Therapy, 17th Ed. Whitehouse Station, NJ: Merck Research Laboratories.
15. TRC Companies, Inc. 2004. TRC Web Site. [Online]. Available: http://www.trcsolutions.com/corporate/home.asp for more information [accessed Dec. 21, 2004].
16. EPA (Environmental Protection Agency). 2002. A Review of the Reference Dose and Reference Concentration Processes. EPA/630/P-02/002F. Risk Assessment Forum, U.S. Environmental Protection Agency, Washington, DC. [Online]. Available: http://cfpub.epa.gov/ncea/cfm/recordisplay.cfm?deid=55365 [accessed August 31, 2004].

Appendix C

Participants at Public Sessions

October 27, 2003 – Washington, DC

Persons who made formal presentations

Paul Anastas, Office of Science and Technology Policy
Jonathan Borak, Yale University
Olga Dominguez, National Aeronautics and Space Administration
William Farland, U.S. Environmental Protection Agency, Office of Research and Development
Karen Guevara, U.S. Department of Energy, Office of Environmental Management
Annie Jarabek, U.S. Environmental Protection Agency, Office of Research and Development
Desmond Lugg, National Aeronautics and Space Administration
David Mattie, U.S. Department of Defense, Air Force Research Laboratory
Dan Rogers, U.S. Department of Defense, Air Force Legal Services Agency
Jennifer Sass, Natural Resources Defense Council
Tom Schneider, Office of Senator Dianne Feinstein
Jonathan Tolman, Senate Committee on Environment and Public Works
Terry Troxell, Food and Drug Administration, Office of Plant and Dairy Foods and Beverages

Persons who made comments at the open microphone session

John Gibbs, Kerr-McGee Shared Services Company
Hank Giclas, Western Growers Association

Larry Ladd, Community Advisory Group for Aerojet Superfund Site Issues, Rancho Cordova, California
Steven Lamm, Consultants in Epidemiology and Occupational Health
Richard Pleus, Intertox

December 12, 2003 – Irvine, CA

Persons who made formal presentations

Rebecca Clewell, CIIT Centers for Health Research
Christopher De Rosa, Agency for Toxic Substances and Disease Registry
Janice Juraska, University of Illinois–Urbana-Champaign
Harold Schwartz, University of California–Irvine
Douglas Wolf, U.S. Environmental Protection Agency, Environmental Carcinogenesis Division

December 13, 2003 – Irvine, CA

Persons who made formal presentations

Jonathan Borak, Yale University
F. Robert Brush, San Diego State University
Patricia Buffler, University of California–Berkeley
Kenneth Crump, Environ
Kirby C. Donnelly, Texas A&M University
Andrea Elberger, University of Tennessee Health Sciences Center
Matthew Hagemann, Soil/Water/Air Protection Enterprise
Steven C. Lewis, ExxonMobil Biomedical Sciences
Richard Pleus, Intertox
Sam Sanderson, University of Nebraska Medical Center
David Ting, California Environmental Protection Agency, Office of Environmental Health Hazard Assessment
Douglas Wahlsten, University of Alberta

Persons who made comments at the open microphone session

William Campbell, Tohono O'Odham Nation Water Resources
Todd Croft, Nevada Division of Environmental Protection
Bill Gedney, Southern California Water Company

Dan Guth, Boeing Company
Gary Praglin, Engstrom, Lipscomb & Lack
Harold Schwartz, University of California–Irvine
Renee Sharp, Environmental Working Group
Lenny Siegel, Center for Public Environmental Oversight
James Strock, Council on Water Quality

May 24, 2004 – Woods Hole, MA

Persons who made formal presentations

Jonathan Borak, Yale University
William Farland, U.S. Environmental Protection Agency, Office of Research and Development
John Gibbs, Kerr-McGee Corporation
Curtis Klaasen, University of Kansas Medical Center
William Mendez, ICF Consulting, Inc.
David Ting, California Environmental Protection Agency, Office of Environmental Health Hazard Assessment

Persons who made comments at the open microphone session

Dan Guth, Boeing Company
Steven Lamm, Consultants in Epidemiology and Occupational Health
James Strock, Council on Water Quality
Carol Rowan West, Massachusetts Department of Environmental Protection

Appendix D

Sensitivity of Perchlorate-Induced Iodide Uptake Inhibition to Serum Iodide Concentrations

THE committee performed calculations on the effect of variations in basal serum iodide concentrations in humans on the perchlorate-induced inhibition of the rate of sodium-iodide symporter (NIS)-mediated uptake of iodide by thyroid cells. The committee used the same Michaelis-Menten competitive inhibition equation and parameters for iodide uptake that were used in the human physiologically based pharmacokinetic models as follows:

$$R_{up}TF_I = \frac{V_{\max}TF_I * CTS_I}{K_mTF_I\underbrace{(1 + \frac{CTS_P}{K_mTF_P})}_{\text{competitive inhibition with } ClO_4^-} + CTS_I},$$

where

$R_{up}TF_I$	=	rate of uptake of iodide from thyroid stroma (capillary bed) to thyroid cells mediated by NIS (in nanograms per hour [ng/hr]),
$V_{max}TF_I$	=	maximal velocity (capacity) for transport of iodide by NIS (1.31×10^6 ng/hr for 70-kg human),
K_mTF_I	=	Michaelis constant (affinity) for iodide transport by NIS (4.0×10^6 nanograms per liter [ng/L]),

K_mTF_P = Michaelis constant (affinity) for perchlorate transport by NIS (1.8×10^5 ng/L),

CTS_I = concentration of iodide in thyroid stroma (capillary bed) (in ng/L, to simulate a range of steady-state concentrations),

CTS_P = concentration of perchlorate in thyroid stroma (capillary bed) (in ng/L, to simulate a range of steady-state concentrations).

The committee calculated the percent inhibition of the iodide uptake induced by perchlorate at various concentrations that span the range of values measured in humans, from a typical basal iodide concentration of 1.5 micrograms per deciliter (µg/dL) to 15 µg/dL. The calculations assume that steady-state thyroid stroma concentrations are the same as serum concentrations of perchlorate and iodide. Results are shown in Figure D-1.

The overlapping curves in Figure D-1 show that variations in basal iodide concentrations have no effect on the competitive inhibition by perchlorate of the rate of iodide uptake by thyroid cells. Thus, one could conclude that serum iodide concentrations (assumed to be the same as thyroid stroma concentrations in these calculations) in humans ranging from 1.5 to 15 µg/dL would not alter the sensitivity of a study performed to evaluate the impact of perchlorate on the determination of iodide uptake by thyroid cells via the NIS.

The committee also ran calculations of ever-increasing basal iodide concentrations to determine the concentrations that would have to be present before a decrease in sensitivity to perchlorate would be observed. These simulations are shown in Figure D-2.

As shown in Figure D-2, a basal iodide concentration over 100 µg/dL would be needed to shift the dose-response curve for the effects of perchlorate on the inhibition of iodide uptake by thyroid cells; that is, concentrations would have to approach the Michaelis constant (affinity) for NIS transport of iodide. Such high concentrations do not seem plausible. Thus, the committee concludes that humans who have serum iodide concentrations of 0-100 µg/dL would be equally sensitive to perchlorate's effects on thyroid iodide uptake (all other things being equal).

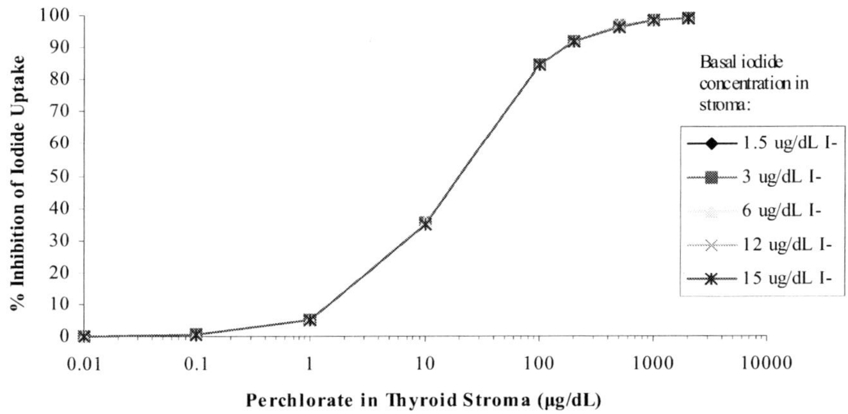

FIGURE D-1 Dose-dependent inhibition of iodide uptake by perchlorate over a range of basal iodide concentrations in thyroid stroma (1.5-15 µg/dL). Abbreviations: I , iodide; µg/dL, micrograms per deciliter.

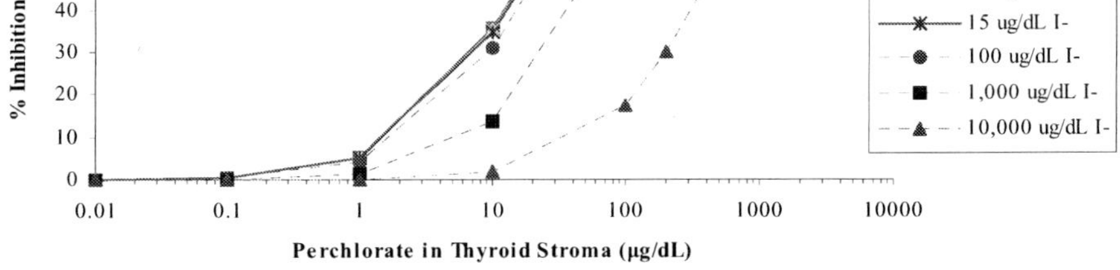

FIGURE D-2 Dose-dependent inhibition of iodide uptake by perchlorate over 10^5-fold range in basal iodide concentrations in thyroid stroma (1.5-10,000 µg/dL). Abbreviations: I-, iodide; µg/dL, micrograms per deciliter.

Appendix E

Physiologically Based Pharmacokinetic Modeling

As discussed in Chapter 4, physiologically based pharmacokinetic (PBPK) modeling is one of the methods of choice for determining human equivalent exposures and adjusting default uncertainty factors associated with the derivation of reference doses (RfDs) and reference concentrations (RfCs) for lifetime human exposures from animal studies (EPA 2002a). Thus, the U.S. Environmental Protection Agency (EPA) relied on a series of PBPK models developed by the Department of Defense (DOD) for perchlorate to facilitate interspecies extrapolations in its draft risk assessment (EPA 2002b,c). The PBPK models were initially developed to describe the disposition (absorption, distribution, metabolism, and elimination) of perchlorate in adult rats (Fisher et al. 2000). As data became available and perchlorate-induced effects observed in animal studies and humans were shown to be mediated by interactions with iodide at the sodium-iodide symporter (NIS) in thyroid tissues, the initial model was expanded to include the disposition of iodide in the body and the inhibition of iodide uptake at the NIS in pregnant rats and fetuses, lactating rats and pups, and adult humans to address dose-response issues associated with potentially sensitive populations (Clewell et al. 2001, 2003a,b; Merrill 2001b, Merrill et al. 2003).

The purpose of the models was to facilitate extrapolation of internal, target-tissue doses from animals to humans, of high to low dose, and across routes of exposure in human health risk assessments. However, the challenges associated with developing quantitative descriptions of the complex interactions and feedback mechanisms within the hypothalamic-pituitary-thyroid (HPT) axis prevented the development of a model that could describe the dynamic interactions between the inhibition of iodide transport by the NIS and thyroid function. Thus, the PBPK models that were developed by DOD focused on improving the ability to predict the kinetics of perchlorate and iodide with respect to their interaction at the NIS as the key event identified by EPA over a range of perchlorate exposure that encompasses toxicity studies in animals, therapeutic uses in humans, and relevant environmental exposures.

EPA used that approach in the derivation of an RfD for perchlorate, following established guidelines for the selection of key events, points of departure, and derivation of appropriate uncertainty factors (EPA 2002a,b,c). Although the committee has chosen a different point of departure based on human data (see Chapter 5), it agrees with EPA that PBPK modeling constitutes the best approach to determining the human equivalent exposures and adjustments to default uncertainty factors when RfDs are based on data collected in animals. If future studies with perchlorate are conducted in laboratory animals, the PBPK models developed to date may be important tools for integrating the new data into the existing database on the exposure-dose-response relationships for perchlorate in rats and humans.

EPA provided a thorough description of the PBPK models in its analysis of the health risks associated with potential perchlorate exposures (EPA 2002b). The models were later reviewed by an external peer-review panel (EPA 2002c). In this appendix, the committee reviews only the general properties of the PBPK models developed by DOD, their underlying assumptions, and their general applicability in animal-to-human extrapolations.

GENERAL APPROACHES TO PERCHLORATE PBPK MODEL DEVELOPMENT

Perchlorate does not appear to be metabolized in the body. Once absorbed, it is rapidly distributed into all tissues except fat, with preferential uptake by tissues that contain the NIS, and ultimately cleared unchanged in urine. Thus, the development of the PBPK models for adult rats and humans and potentially sensitive life stages (developing fetus and neonate) followed a logical progression of increasing complexity that linked exposure with key biochemical events as the toxicity and mode of action of perchlorate in animals and humans became better defined and methods of analyzing perchlorate in biologic fluids and tissues improved.

In deriving the RfD, EPA relied on the initial published models and a series of memoranda from DOD describing the continued updating of the suite of PBPK models (Clewell 2001a,b; Merrill 2000; Merrill 2001a,b). The final PBPK models were published shortly after the EPA review (Clewell et al. 2003a,b; Merrill et al. 2003) and reflected comments received by EPA's peer review (EPA 2002c) and journal peer reviewers. The models have similar base structures and sources for model parameter values and additional features where necessary to address specific life stages, such as growth and development and additional compartments associated with pregnancy and lactation in the rat.

The first PBPK model of Fisher et al. (2000) was developed to describe the disposition of perchlorate in the adult male rat; it was based on rather sparse kinetic data that were available in the late 1990s. Tissues that were specifically included in the initial framework included lungs, kidneys, thyroid, and gastrointestinal (GI) tract; the remaining tissues were lumped into either poorly perfused or richly perfused compartments (Figure E-1). Each of those compartments consisted of mass-balance equations describing the rates of transfer of perchlorate into and out of each tissue on the basis of their known volumes, blood perfusion rates, partition coefficients, the presence (or absence) of any transport processes (such as that of the NIS in thyroid and GI tract), or the clearance of perchlorate into urine. Only sparse data on perchlorate after intravenous (IV) administration were available for model development. On the basis of the the preliminary analysis, it was observed that perchlorate clearance into urine after IV dosing at 0.01-3 mg/kg followed linear, first-order kinetics, whereas substantial nonlinearities were observed in systemic (serum and tissue) kinetics. Thus, the preliminary model was used as an initial framework for identifying critical data gaps for later model development and definitions of relevant internal-dose surrogates (as opposed to administered doses or external exposures) that could be used to enhance human health risk assessments that were based on toxicity studies in rats.

Clewell et al. (2001) later proposed a suite of initial PBPK models that incorporated important steps in the mode of action of perchlorate (NIS iodide uptake inhibition) and key life stages for perchlorate toxicity (fetus and neonate) in addition to the adult male. To accomplish that, initial PBPK models were developed for perchlorate and iodide with interactions between them occurring at the NIS. The purpose of the preliminary PBPK models was to lay the groundwork for extrapolating the internal doses of perchlorate (such as blood concentrations and interactions with iodide in thyroid NIS) from rats to humans as a function of life stage (fetus, neonate,

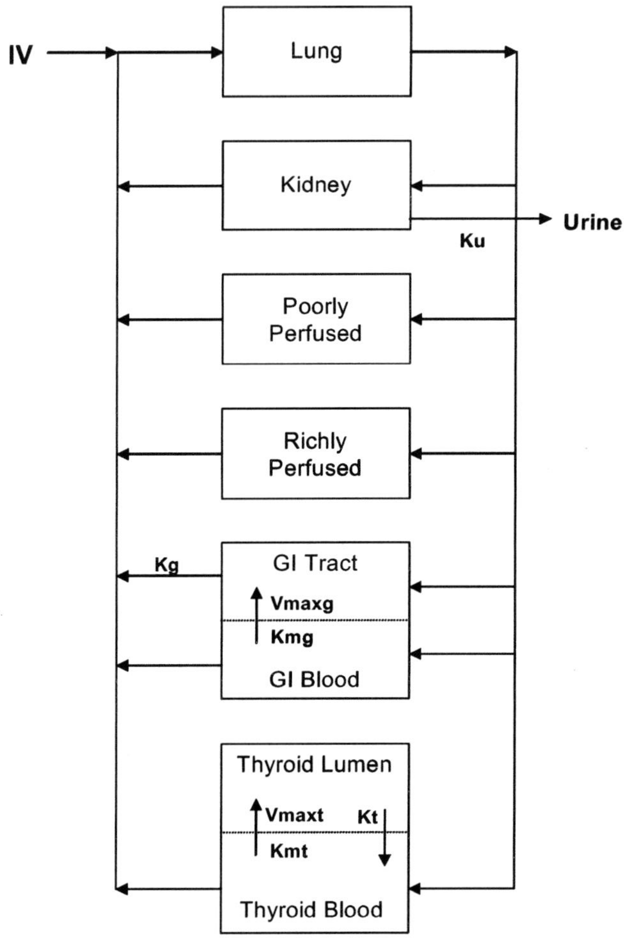

FIGURE E-1 Diagram of preliminary PBPK model for perchlorate in adult male rat. Abbreviations: GI, gastrointestinal; IV, intravenous; K_g, rate constant for loss of perchlorate from GI tract; K_{mg}, rate constant for uptake of perchlorate in GI tract; K_{mt}, rate constant for uptake of perchlorate in thyroid; K_t, rate constant for loss of perchlorate in thyroid; K_u, rate constant for urinary excretion of perchlorate; V_{maxg}, maximal capacity for NIS transport in GI tract; V_{maxt}, maximum capacity for NIS transport in thyroid. Source: Adapted from Fisher et al. 2000.

and adult), dose, and route of exposure with the parallelogram approach discussed in Chapter 4.

The structures of the PBPK models across life stages shared physiologic compartments and biologic and chemical-specific sources of parameter values. Each model contained specific descriptions of tissues involved in the uptake, distribution, and elimination of perchlorate and iodide and specific descriptions of the interactions between perchlorate and iodide via the NIS. Basic model structures thus included NIS-containing tissues, such as the thyroid, GI tract, and skin. The thyroid was divided into three - subcompartments representing the tissue capillary bed (referred to as the stroma in the perchlorate PBPK literature), follicle, and lumen (colloid), as described in more detail below. The GI tract was similarly divided into three subcompartments: the capillary bed, GI tissue, and GI contents; perchlorate and iodide are transported by the NIS from the tissue into the contents against a concentration gradient and diffuse between subcompartments on the basis of partition coefficients and electrochemical gradients. The skin was divided into two subcompartments: skin blood and skin tissue; perchlorate and iodide are transported from blood to tissue by the NIS and diffuse back and forth between the subcompartments. To model pregnancy and lactation, NIS-containing tissues, such as the placenta and mammary glands, were included. Those tissues were also divided into subcompartments as described below for each model. Other NIS-containing tissues—such as ovaries, choroids plexus, and salivary glands—were considered too small to affect the concentrations of perchlorate and iodide in plasma or the thyroid and were therefore not specifically segregated from the poorly or richly perfused tissue groups.

Remaining tissues in each model included the kidneys for clearance of perchlorate and iodide in urine, the liver for possible future modeling of hormone metabolism, a separate plasma and red-cell compartment for the distribution of perchlorate (plasma-protein-binding) and iodide, and fat because it is an exclusionary compartment (owing to the poor solubility of perchlorate and iodide anions) that is highly variable and changes during pregnancy and lactation. The lung compartment used in the preliminary model of Fisher et al. (2000) was combined with the richly perfused compartment because neither perchlorate nor iodide is volatile; vapor inhalation was considered to be an irrelevant route of exposure, and neither anion is eliminated to any important degree in breath. Remaining tissues were grouped together on the basis of their similar kinetic properties and blood perfusion rates (as poorly vs richly perfused tissues).

As the authors of the PBPK models discussed, it is important to note that the iodide PBPK models used for each life stage (adult, pregnant, lactating, fetus, or neonate) were rudimentary in that they described key physiologic and biochemical processes only in enough detail to reproduce the sparse radiolabeled-iodide kinetic data in animals and humans and the interactions with perchlorate at the NIS. With one exception, discussed below, the models did not include dietary or endogenous iodide kinetics, which could be important in interpreting human biomonitoring or clinical studies. And the models did not include formal descriptions of the pharmacodynamics of thyroid hormone control and the HPT biofeedback effects of alterations in iodide concentrations resulting from perchlorate exposure (although NIS up-regulation was modeled to fit the iodide kinetic data after multiple dosing kinetic studies with perchlorate, as discussed below). Such a combined PBPK-pharmacodynamic model would eventually be useful in predicting biologic responses (alterations in thyroid hormone homeostasis) to various perchlorate exposures or different life stages, as noted in EPA (2002c), and in providing a more quantitative basis for understanding species differences in potential toxicity.

MODELING KEY EVENTS IN NIS-MEDIATED THYROID RESPONSES TO PERCHLORATE

As discussed in Chapter 5, the committee considers the inhibition of iodide uptake at the NIS in the thyroid as one of the key biochemical events preceding the development of adverse responses, such as hypothyroidism, which may result in abnormal growth and development of the fetus and child or in metabolic sequelae at any age. It is important to emphasize that the inhibition of thyroid iodide uptake is not in itself an adverse response. However, because the inhibition of iodide uptake is the first definitive biochemical event that must occur before the chain of events leading to altered hormone homeostasis and then an adverse response, it is a critical component in establishing a relationship between exposure and an internal dose-response for perchlorate risk assessments. Thus, the PBPK models used by EPA in the derivation of an RfD for perchlorate focused on adequately describing the interaction between perchlorate, iodide, and the NIS in the thyroid gland even though the internal-dose metric used by EPA in calculating human equivalent exposures was limited to the blood perchlorate concentration.

The working hypothesis underlying the development of the PBPK models for perchlorate and iodide across life stages of concern was that perchlorate competitively inhibits the thyroid uptake of iodide via the NIS and reduces the iodide available for the production of thyroid hormones (Figure E-2). The inhibition of iodide uptake results in decreases in the production of triiodothyronine (T_3) and thyroxine (T_4), which lead to an increase in thyroid-stimulating hormone (TSH) production that then stimu-

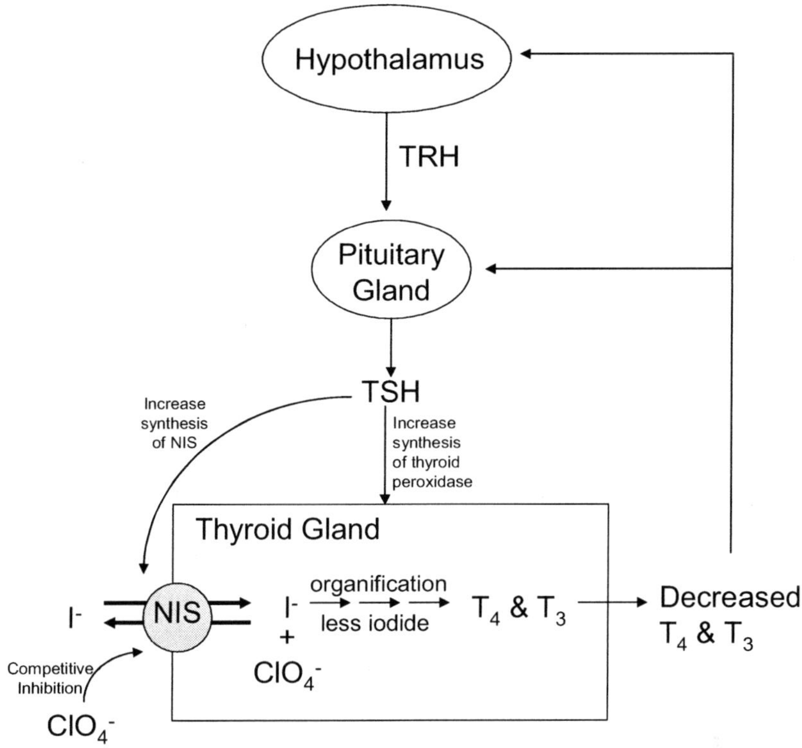

FIGURE E-2 Working hypothesis on mode of action of perchlorate on thyroid gland. Abbreviations: ClO_4^-, perchlorate ion; I^-, iodide; NIS, sodium-iodide symporter; T_3, triiodothyronine; T_4, thyroxine; TRH, thyrotropin-releasing hormone; TSH, thyroid-stimulating hormone.

lates an increase in the synthesis of NIS proteins and enzymes associated with hormone production.

As reviewed in Chapter 4, there are important species differences in the distribution and disposition of thyroid hormones. In particular, the reduced protein binding of T_3 and T_4 in rat serum (rats lack a thyroxine-binding globulin that is normally present in humans) results in a greater rate of clearance of thyroid hormones from the body than in humans. That causes a compensatory increase in TSH release from the pituitary gland and a corresponding increase in the overall production rate of T_3 and T_4, making rats particularly sensitive to perchlorate's disruptions in iodide uptake.

During the first 10-12 weeks of gestation, the fetus depends on the mother for thyroid hormone production, so the fetus is affected by both alterations in the mother's hormone production and direct inhibition of iodide uptake by perchlorate that has transferred across the placenta. One of the major goals in the development of the PBPK models of Clewell et al. (2003a,b), Merrill (2001), and Merrill et al. (2003) was to simulate interactions between perchlorate and iodide at the NIS in the thyroid of rats and humans at key life stages (adult, pregnancy, lactation, fetus, and neonate) and thereby to facilitate high-to-low dose, route-to-route, and cross-species extrapolations based on an internal dose of perchlorate that is relevant to toxicity in both cancer and noncancer risk assessments. To accomplish that goal, a PBPK model for iodide was also developed that interfaced with the perchlorate PBPK model at the level of the NIS in the thyroid gland, GI tract, and skin.

A fundamental component of the assumed competitive interaction between perchlorate and iodide is transport of perchlorate itself by the NIS. There has been some debate regarding that issue (Wolf 1998; Reidel et al. 2001a,b; EPA 2002c), but it has not been possible to measure perchlorate uptake directly in rat and human thyroid follicular cells, because of a lack of commercially available radiolabeled perchlorate or an analytic method that is sensitive enough. Thus, arguments have largely centered on the use and interpretation of indirect estimates of perchlorate transport (such as electrochemical gradient measurements vs measurement of radiolabeled uptake of chemicals similar to perchlorate in cells that contain the NIS). Until data become available to address the question directly, the committee agrees that the use of a competitive-inhibition model and its underlying assumption of perchlorate transport by the NIS, as reviewed by Clewell et al. (2004), is a reasonable approach and provides an adequate description of the available data.

To model the interactions between perchlorate and iodide, Clewell et al. (2003a,b), Merrill (2001a,b), and Merrill et al. (2003) reduced the complexity in potential biologic interactions and maintenance of thyroid hormone homeostasis to a series of rate-limiting steps in a physiologically based compartment specific to each species or life stage (adult human, adult male rat, pregnant female rat and fetus, and lactating female rat and neonate). In their simplified models, the thyroid was divided into three compartments representing the capillary bed (stroma), follicle cells, and colloid (lumen) after less complex descriptions (two compartments) failed to simulate the initial rapid phase of thyroid perchlorate uptake and equilibrium and a slower phase of equilibrium and clearance observed in animal studies (Figure E-3).

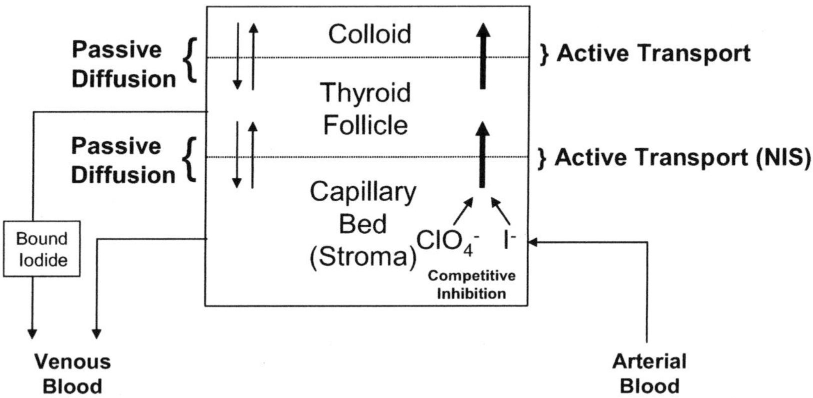

FIGURE E-3 Diagram of three-compartment working model for perchlorate and iodide uptake in the thyroid. Abbreviations: ClO_4^-, perchlorate ion; I^-, iodide; NIS, sodium-iodide symporter. Source: Adapted from Clewell et al. 2004.

In the simplified model, both perchlorate and iodide were allowed to partition from arterial blood into the thyroid capillary bed on the basis of species-specific and life-stage-specific blood flow rates, tissue:blood partition coefficients, and tissue volumes. The physiologic model thus consisted of a series of mass-balance equations that described the overall rate of change in the amount of perchlorate or iodide in each tissue or tissue compartment as a function of the rate of chemical input less the rate of chemical output:

$$\begin{array}{l} \text{Rate of Change} \\ \text{of Chemical} \\ \text{in Tissue “i”} \end{array} = \begin{array}{l} \text{Rate of Input} \\ \text{of Chemical} \\ \text{in Tissue “i”} \end{array} - \begin{array}{l} \text{Rate of Output} \\ \text{of Chemical} \\ \text{in Tissue “i”.} \end{array} \qquad (1)$$

For the thyroid capillary bed compartment—designated as "stroma" in Clewell et al. (2003a,b), Merrill (2001a,b), and Merrill et al. (2003) model equations described below—perchlorate and iodide that enter from the arterial blood can either passively diffuse or be actively transported by the NIS into the thyroid follicle and passively diffuse back from the follicle or partition back into venous blood draining the thyroid. Thus, the following mass-balance equation was used by Clewell et al. (2003a,b), Merrill (2001a,b), and Merrill et al. (2003) to represent the rate of change in the amount of perchlorate in the thyroid stroma ($RATS_P$) according to

$$RATS_P = \underbrace{QT(CA_P - CVTS_P)}_{\substack{\text{partitioning} \\ \text{with blood}}} + \underbrace{PATF_P\left(\frac{CTF_P}{PTF_P} - CTS_P\right)}_{\substack{\text{passive diffusion} \\ \text{follicle to stroma}}} - \underbrace{RupTF_P}_{\substack{\text{NIS transport} \\ \text{stroma to follicle}}}, \qquad (2)$$

where QT is the thyroid blood flow, CA_P is the concentration of perchlorate in arterial blood, $CVTS_P$ is the concentration of perchlorate in venous blood draining the thyroid stroma, $PATF_P$ is the permeability:area cross product for the diffusion of perchlorate across the membrane between the stroma and follicle, CTF_P is concentration of perchlorate in thyroid follicle, PTF_P is the thyroid stroma:follicle partition coefficient for perchlorate, CTS_P is concentration of perchlorate in thyroid stroma, and $RupTF_P$ is the rate of active transport of perchlorate by the NIS from stroma to follicle. Identical equations were used to describe the rate of change in the concentration of iodide in the thyroid stroma with iodide-specific chemical parameters, such as the concentration of iodide in arterial blood, thyroid stroma, and thyroid follicle; permeability-area cross products for the diffusion of iodide across the thyroid follicle membrane; the thyroid stroma:follicle partition coefficient for iodide; and the rate constants associated with the active transport of iodide by NIS.

The rate of uptake of perchlorate into the thyroid follicle ($RATF_P$) was then described by

$$RATF_P = \underbrace{RupTF_P}_{\substack{\text{NIS} \\ \text{transport} \\ \text{stroma to follicle}}} + \underbrace{PATF_P\left(CTS_P - \frac{CTF_P}{PTF_P}\right)}_{\substack{\text{passive diffusion} \\ \text{stroma to follicle}}} - \underbrace{RupTL_P}_{\substack{\text{active} \\ \text{transport} \\ \text{follicle to colloid}}} + \underbrace{PATL_P\left(\frac{CTL_P}{PTL_P} - CTF_P\right)}_{\substack{\text{passive diffusion} \\ \text{follicle to colloid}}}, \qquad (3)$$

where $RupTL_P$ is the rate of active transport of perchlorate from follicle into colloid (lumen), $PATL_P$ is the permeability:area cross product for the diffusion of perchlorate across the membrane between follicle and colloid, CTL_P is the concentration of perchlorate in thyroid colloid, PTL_P is the thyroid follicle:colloid partition coefficient for perchlorate, and CTF_P is the concentration of perchlorate in thyroid follicle. For iodide, the same equation (Equation 3) was used with the addition of terms describing the loss of inorganic iodide due to the production of thyroid hormones (see "Bound

Iodide," Figure E-3) as a simple, first-order reaction lumping total hormone production:

$$(\text{Equation 3}) - (CLProd_I)(CTF_I), \tag{4}$$

where $CLProd_I$ is the first-order rate of organification of iodide and CTF_I is the concentration of inorganic iodide in the thyroid follicle.

The rate of change in the amount of perchlorate in the thyroid colloid (lumen) then becomes

$$RATL_P = \underbrace{RupTL_P}_{\substack{\text{active}\\ \text{transport}\\ \text{follicle to colloid}}} + \underbrace{PATL_P(CTF_P - \frac{CTL_P}{PTL_P})}_{\substack{\text{passive diffusion}\\ \text{follicle to colloid}}}\ , \tag{5}$$

The active transport of perchlorate from the thyroid stroma to the follicle ($RupTF_P$) and from the follicle to the lumen ($RupTL_P$) was described by using Michaelis-Menten equations as the basis of saturable transport processes according to

$$RupTF_P = \frac{V_{max}TF_P * CTS_P}{K_mT_P + CTS_P} \tag{6}$$

and

$$RupTL_P = \frac{V_{max}TL_P * CTF_P}{K_mTL_P + CTF_P}\ , \tag{7}$$

where $V_{max}TF_P$ and $V_{max}TL_P$ are the maximal rates of transport from stroma to follicle via the NIS and follicle to lumen via other transporters (such as pendrin or apical iodide channels), respectively, and K_mT_P and K_mTL_P are the Michaelis constants for the transport of perchlorate from stroma to follicle and follicle to lumen, respectively.

For the integration between the perchlorate and iodide models, the basic Michaelis-Menten equations for NIS transport (stroma to follicle) and active transport in the apical membrane (follicle to colloid) were modified by Clewell et al. (2003a,b), Merrill (2001a,b), and Merrill et al. (2003) to incorporate the competitive inhibition of iodide uptake by perchlorate, as follows:

$$RupTF_I = \frac{V_{\max}TF_I * CTS_I}{\underbrace{KmTF_I(1+\frac{CTS_P}{KmTF_P})}_{\text{competitive inhibition with } ClO_4^-} + CTS_I}, \quad (8)$$

and

$$RupTL_I = \frac{V_{\max}TL_I * CTF_I}{\underbrace{KmTL_I(1+\frac{CTF_P}{KmTL_P})}_{\text{competitive inhibition with } ClO_4^-} + CTF_I}, \quad (9)$$

where the rates of transport of perchlorate and iodide by the NIS and apical membrane transporter are mediated by the concentrations of each chemical in the stroma (CTS_P and CTS_I) or follicle (CTF_P and CTFI) and the respective Michaelis constants (K_mTF_P, K_mTF_I, K_mTL_P, and K_mTL_I). Because of the low Michaelis constants for NIS-based active transport of perchlorate ($K_mTF_P = 1\text{-}1.8 \times 10^5$ ng/L) versus iodide ($K_mTF_I = 4.0 \times 10^6$ ng/L) and for apical membrane transport of perchlorate ($K_mTL_P = 1 \times 10^8$) versus iodide ($K_mTL_I = 1 \times 10^9$), iodide has little effect on perchlorate kinetics. Thus, Merrill (2001a,b), Merrill et al. (2003), and Clewell et al. (2003a,b) did not include the corresponding competitive inhibition of perchlorate transport by iodide in their perchlorate models. Although the committee considers the inclusion of the competitive inhibition of thyroid perchlorate uptake by iodide as being more consistent with the underlying assumption of competitive inhibition, it agrees with the authors that this simplification has little or no effect on the PBPK simulations.

Thus, the biochemical parameters controlling the uptake of perchlorate into the thyroid gland and within the three main regions (capillary bed or stroma, follicle, and colloid or lumen) include the tissue:blood partition coefficient; permeability-area cross products for passive diffusion between thyroid regions; the NIS between capillary bed and follicle, where perchlorate and iodide compete for transport; and a second apical membrane active transport process between follicle and colloid regions, where perchlorate also competitively inhibits iodide uptake.

ADULT RAT PBPK MODEL

Merrill et al. (2003) extended the preliminary PBPK model of Fisher et al. (2000) to provide a more detailed description of the disposition of perchlorate and its interactions with iodide in adult male rats. The model included oral gavage, IV injection, and drinking-water routes of exposure; target tissues (thyroid); tissues important to the distribution of perchlorate (GI tract, liver, kidney, skin, fat, blood, and remaining richly and poorly perfused tissues as lumped compartments); and elimination of perchlorate in urine, as shown in Figure E-4.

The anatomic and physiologic structure of the iodide PBPK model is identical with that of the perchlorate model except for the blood and thyroid compartments. In the thyroid, inorganic iodide is used in the production of thyroid hormones (separate compartment representing total thyroid hormone-base iodide designated as "bound iodide" in the model structure [Figure E-4]), whereas perchlorate is nonreactive and, like inorganic iodide, can passively diffuse from the thyroid lumen and follicles back into the thyroid stroma, where it can partition into venous blood draining the thyroid tissues. This "bound iodide" compartment was necessary to describe the radioiodide kinetic studies (which do not differentiate the form of the radiolabel) because both free (inorganic) and bound (organic) iodide is taken up by tissues. Thus, the bound iodide was pooled with the inorganic iodide in the blood compartment for distribution of total iodide in the body. That simplification assumes that radioiodide behaved as inorganic iodide in all compartments other than the thyroid, including serum. In the thyroid, first-order rate constants were used to describe the formation of bound iodide and its later release into systemic circulation.

An additional difference in the structure of the blood compartment was the need to explicitly include a description of protein binding of perchlorate in plasma but not iodide (iodide in blood already represented inorganic and organic iodide). Michaelis-Menten equations were used in the blood compartment (similar to the format used in Equations 6 and 7) to describe the association between free perchlorate and an unspecified plasma-protein (presumably albumin) binding site. Although such a simple description has been used to describe plasma-protein binding for many other chemicals, Merrill et al. (2003) also included a first-order rate constant for the dissociation of perchlorate from the binding protein(s). It is not clear to the committee why the extra parameter was needed in the model.

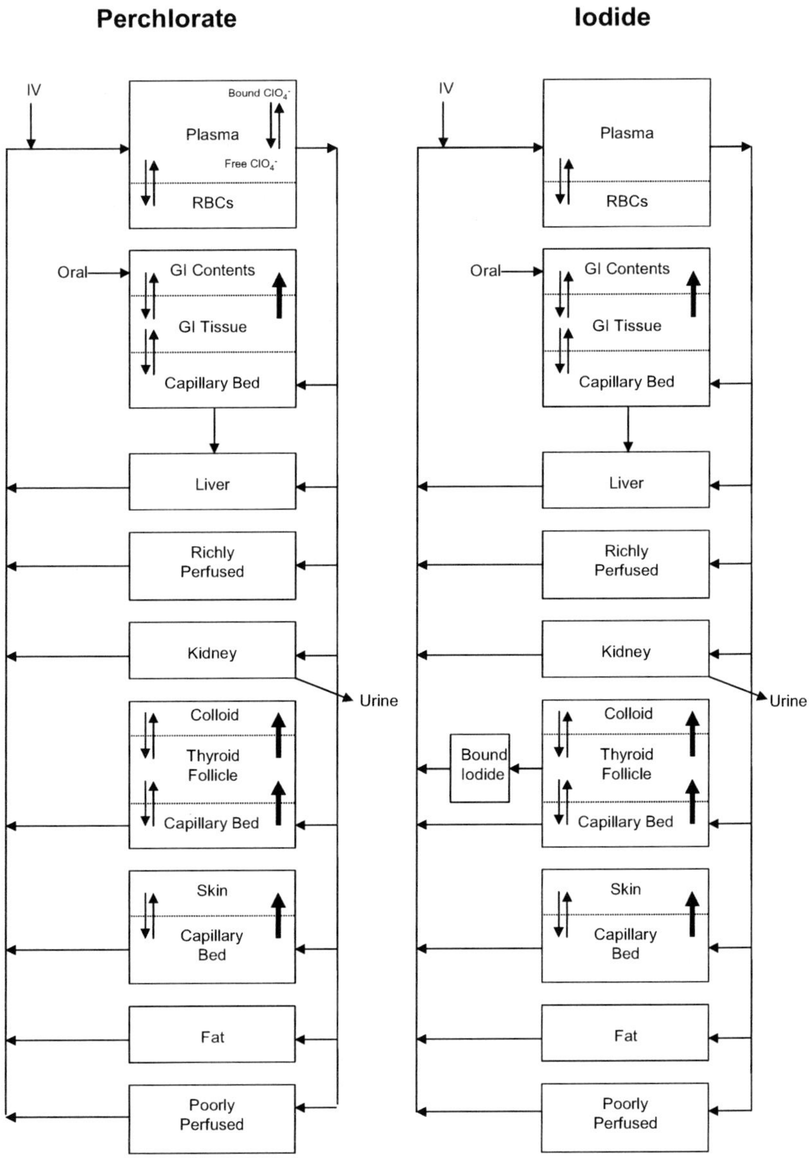

FIGURE E-4 Diagram of PBPK models of perchlorate and iodide in adult rat. Bold arrows indicate active transport of perchlorate and iodide NIS in thyroid, skin, and gastric mucosa and by apical iodide channels in thyroid. Abbreviations: ClO_4^-, perchlorate ion; GI, gastrointestinal; IV, intravenous; NIS, sodium-iodide symporter; PBPK, physiologically based pharmacokinetic; RBCs, red blood cells. Source: Adapted from Merrill et al. 2003.

Both Merrill et al. (2003) and EPA (2002b) provided a detailed description of the sources for each of the physiologic and chemical-specific model parameter values and the ability of the model to describe a variety of datasets. The basis of those model parameter values has been extensively peer-reviewed, so only selected parameters and simulation issues will be highlighted here. For instance, increases in thyroid perchlorate were observed in 2-week drinking-water exposure studies conducted by Yu et al. (2002) and Merrill et al. (2003) as a function of dose and duration of exposure (for example, at higher dose levels, thyroid perchlorate concentrations were significantly greater after 2 weeks of exposure than after a single dose). In the original PBPK model version summarized by EPA (2002c), such repeated-exposure (vs acute) studies were simulated by increasing the effective perchlorate tissue-partition coefficient between thyroid follicle and thyroid stroma from 0.13 at the lower drinking-water dose levels (0.01, 0.1, and 1.0 mg/kg per day) to 0.4, 1.25, and 2.0 at 3, 10, and 30 mg/kg per day, respectively. There is little biologic reason to justify a change in tissue partitioning as a function of perchlorate dose just to fit the observed data. Therefore, in the final publication, Merrill et al. (2003) presented a more biologically consistent alteration in the model to describe the nonlinear behavior in perchlorate kinetics as a function of dose and duration of exposure. The alteration consisted of increasing the maximum capacity for NIS transport, which is known to be up-regulated by increased TSH, rather than changing the tissue partitioning.

No direct measures of the maximum capacity for NIS transport were available, so the maximum capacity for NIS transport in the thyroid (the only tissue where the NIS is up-regulated by TSH) was increased until predicted thyroid perchlorate and iodide corresponded with values measured for each dose used in the drinking-water studies. The resulting maximum capacity for NIS transport was plotted against the corresponding free (unbound) serum perchlorate predicted by the model at each dose (although the authors did not specify the timing of the serum concentrations), and a Michaelis-Menten equation was derived from the resulting plot. The equation was then used as a surrogate of up-regulation of thyroid NIS transport as a function of free perchlorate in serum to simulate repeated-dosing, drinking-water exposure scenarios.

Such an adjustment of the model worked well for describing the kinetics of perchlorate in the thyroid when up-regulation reached a steady state but does not describe the time lag that occurs during the transition from basal to an up-regulated state. In essence, the PBPK model used either a basal transport level or an up-regulated level in its simulations without a transition phase. The lag for up-regulation could be an important difference between rats and humans that may warrant future studies because of the greater hormone production rates and low storage capacity in rats. Nevertheless, the committee considers the approach used by Merrill et al. (2003) and later in the other PBPK models that simulated both acute and repeated dosing kinetics as a reasonable, biologically consistent simplification for describing the up-regulation of NIS transport in the thyroid. Furthermore, such up-regulation has little effect on serum concentrations of perchlorate, which were used as a basis of human equivalent exposure calculations in the derivation of the RfD for perchlorate.

Merrill et al. (2003) performed a sensitivity analysis to determine which model parameters had the greatest effect on the simulations of serum perchlorate concentrations (measured as area under the curve) at drinking-water concentrations above (10 mg/kg per day) and below (0.1 mg/kg per day) saturation of the NIS. The results of the analysis indicated that chemical-specific parameters associated with plasma-protein binding of perchlorate, uptake of perchlorate in the skin, and urinary clearance of perchlorate had the greatest influence on serum perchlorate simulations. Thus, it is particularly important to have independent data on those parameters.

Merrill et al. (2003) optimized the perchlorate plasma-protein binding parameter values from in vivo kinetic data, such as Yu et al. (2002). In vitro equilibrium dialysis studies in rat plasma were available to confirm the parameter values. For urinary clearance of perchlorate, Merrill et al. (2003) likewise optimized a first-order clearance parameter value from available time-course kinetic data. The first-order clearance value, 0.07 L/hr per kilogram of body weight, is about 25% of the glomerular filtration rate for a male Sprague-Dawley rat, which supports the need to include protein binding in the model in that perchlorate should be readily filtered by the glomeruli. Given the importance of urinary clearance for the model to simulate serum perchlorate concentrations, and thus inhibition of thyroid iodide uptake, independent studies should be conducted to improve the estimation of urinary clearance of perchlorate (and iodide) in future models.

Owing to its large size (19% of the body weight of the rat) and the presence of the NIS, the skin compartment was also identified in sensitivity analyses as an important tissue that affects the kinetics of both perchlorate and iodide. Yu et al. (2002) and Merrill et al. (2003) included skin in their pharmacokinetic studies with rats, but there appeared to be considerable variability in data in rats and a paucity of data in humans regarding the localization of perchlorate and iodide in skin, as discussed below. However, the skin compartment does not represent as great a percentage of the body weight of humans as of rats (19% vs about 3.7%). Thus, it may be more important to develop a thorough understanding of the kinetics of the two anions in rat skin than in humans.

Although occasional datasets were not particularly well described, the committee agrees with EPA that the male rat PBPK model of Merrill et al. (2003) provided a reasonable description of a variety of datasets in rats after IV, subcutaneous, or intraperitoneal (IP) injection of drinking-water exposure to perchlorate and oral gavage or IV and IP injection of iodide. To facilitate comparisons among the final PBPK models, the final physiologic and biochemical values for the adult rat PBPK model and the other PBPK models of the pregnant rat and fetus, lactating rat and neonate, and adult human are summarized together in Table E-1.

ADULT HUMAN PBPK MODEL

EPA relied on a memorandum from Merrill (2001b) as the basis of human PBPK-model simulations for perchlorate and iodide. The structure of the adult human PBPK model was nearly identical with that of the male rat model of Merrill at al. (2003) from which it was derived (see Figure E-4) and will therefore not be reiterated here. Male and female human physiology constants were taken from data available in the literature and are summarized in Table E-1. Chemical-specific parameter values were developed in a way analogous to that for the rat model; this often involved fitting of model parameters to available kinetic data on perchlorate and iodide at various dose levels. The parameter values are also summarized in Table E-1. EPA provided a thorough review of the sources of the model parameters; only the ones that are different from those for adult rat are highlighted in this appendix.

For instance, slightly different partition coefficients and permeability-area cross products governing the diffusion of perchlorate and iodide were used for humans in tissues containing the NIS (stomach, skin, GI tract, and thyroid). Such differences are considered minor and probably reflected attempts to refine the human model on the basis of available kinetic data, inasmuch as the human model was being developed at the same time as the adult rat model. Similarly, several parameters governing the transport of perchlorate and iodide by the NIS were also adjusted from the rat model to improve the fits to the available human data. The binding affinity for the NIS appears to be similar across tissues and species, so values for each NIS-

TABLE E-1 Values of Physiologic and Biochemical Parameters for Disposition of Perchlorate and Iodide in Thyroid Compartments of Rat and Human PBPK Models of Clewell (2003a,b), Merrill (2001b), and Merrill et al. (2003)

		Pregnancy (Rat, GD 0-21)[a]		Lactation (Rat, PND 0-18)		Adult
Parameter	Male Rat	Dam	Fetus	Dam	Neonate	Human
Physiologic Parameters for Perchlorate and Iodide Thyroid Compartments						
Body weight (kg)	0.3	0.28-0.361	0-0.0045	0.277-0.310	0.0075-0.1985	~70.0
Tissues volumes						
Slowly perfused (% BW)	74.6	74.6	74.6	37.07-40.42	53.92-49.31	65.1
Rapidly perfused (% BW)	11.0	11.0	16	5.35	5.36	12.4
Fat (% BW)	7.4	10.0-11.0	0	12.45-6.9	0-4.61	21.0(M) 32.7(F)
Kidney (% BW)	1.7	1.7	0.3-0.44	1.7	1.7	0.44
Liver (% BW)	5.5	3.4	8.5-7.2	3.4	3.4	2.6
GI tissue (% BW; VGI)	0.54[b]	3.6	2.0-3.0	3.9	3.9	1.7[b]
GI contents (% BW)	1.68[b]	7.2	0.8-6.2	7.2	7.2	0.071[b]
GI blood (% VGI)	4.1[b]	2.9	2.9	2.9	2.9	4.1[b]
Skin Tissue (% BW; VSk)	19.0	19.0	8.8-19.3	19.0	19.0	3.7
Skin blood (% VSk)	2.0	2.0	2.0	2.0	2.0	8.0
Plasma (% BW)	4.1	4.7	4.7	4.7	4.7	4.4
Red blood cells (% BW)	3.3	2.74	2.74	2.74	2.74	3.5
Placenta (% BW)	—	0-2.57	—	—	—	—
Mammary gland (% BW)	—	1.0-5.5	—	4.4-6.6	—	—
Mammary blood (% mammary)	—	—	—	18.1	—	—
Milk (L)	—	—	—	0.002	—	—
Thyroid total (% BW)	0.0077	0.0105	0.058-0.038	0.0105	0.0125	0.03
Thyroid follicle (% thyroid)	59.9	45.9	61.4	45.89	61.4-37.2	57.3

Thyroid colloid (% thyroid)	24.4	45.0	18.3	45.0	18.3-32.5	15.0
Thyroid blood (% thyroid)	15.7	9.1	20.3	9.1	20.3-30.3	27.6
Cardiac output (L/hr per kg)	14.0	14.0	67.8	14.0-21.0	14.0	16.5
Blood flows (% cardiac output)						
Poorly perfused	24.0	24.0	24.0	7.9-1.9	16.9	13.0
Richly perfused	76.0	76.0	76.0	40.8	40.8	33.0
Fat	6.9	7-8.1	—	7.0	7.0	5.2
Kidney	14.0	14.0	3.6	14.0	14.0	17.5
Liver	17.0	18.0	4.5	18.0	18.0	22.0
GI	1.61[b]	13.6	4.6	1.61	1.61	1.0[b]
Skin	5.8	5.8	10.4	0.058	0.058	5.8[c]
Placenta	—	0.0-12.3	—	—	—	—
Mammary	—	0.2-1.2	—	9.0-15.0	—	—
Thyroid	1.6	1.6	1.6	1.6	1.6	1.6
Perchlorate-Specific Parameters						
Partition coefficients (unitless)						
Slowly perfused:plasma	0 31	0.31	0.31	0.31	0.31	0.31
Rapidly perfused:plasma	0 56	0.56	0.56	0.50	0.50	0.56
Fat:plasma	0 05	0.05	—	0.05	0.05	0.05
Kidney:plasma	0.99	0.99	0.99	0.99	0.99	0.99
Liver:plasma	0.56	0.56	0.56	0.56	0.56	0.56
Gastric tissue:gastric blood	0.70[b]	0.50	1.80	1.80	3.21	1.80[b]
Gastric contents:gastric tissue	1.70[b]	1.30	2.30	2.30	5.64	2.30[b]
Skin tissue:skin blood	1.00	1.15	1.15	1.15	1.15	1.15
Red blood cells:plasma	0.73	0.73	0.73	0.73	0.73	0.80
Placenta:plasma	—	0.56	—	—	—	—

TABLE E-1 *(Continued)*

Parameter	Male Rat	Pregnancy (Rat, GD 0-21)[a] Dam	Pregnancy (Rat, GD 0-21)[a] Fetus	Lactation (Rat, PND 0-18) Dam	Lactation (Rat, PND 0-18) Neonate	Adult Human
Mammary gland:plasma	—	0.66	—	—	—	—
Mammary tissue:mammary blood	—	—	—	0.66	—	—
Mammary tissue:milk	—	—	—	2.39	—	—
Thyroid follicle:stroma	0.15	0.15	0.15	0.13	0.13	0.13
Thyroid lumen:follicle	8.00	7.00	7.00	7.00	7.00	7.00
Permeability area cross products (L/hr per kg)						
Gastric blood-gastric tissue	1.00[b]	1.00	1.00	1.00	1.00	0.60[b]
Perchlorate-Specific Parameters						
Gastric tissue-gastric contents	0.80[b]	1.00	1.00	1.00	1.00	0.80[b]
Skin blood-skin tissue	0.80	1.00	1.00	0.50	1.00	1.00
Plasma-red blood cells	1.00	1.00	1.00	1.00	1.00	1.00
Mammary blood-mammary tissue	—	0.04	—	0.01	—	—
Mammary tissue-milk	—	—	—	0.10	—	—
Placenta blood-placenta tissue	—	0.10	—	—	—	—
Thyroid follicle-stroma	6.0×10^{-5}	6.0×10^{-5}	6.0×10^{-5}	4.0×10^{-5}	4.0×10^{-5}	1.0×10^{-4}
Thyroid lumen-follicle	0.01	0.01	0.01	0.01	0.01	0.01
Thyroid: NIS active transport						
K_m (ng/L)	1.8×10^{5}	1.0×10^{5}	1.0×10^{5}	1.5×10^{5}	1.5×10^{5}	1.8×10^{5}
V_{max} (ng/hr per kg)[d]	1.0×10^{3}	2.6×10^{3}	$0\text{-}2.25 \times 10^{3}$	1.5×10^{3}	1.5×10^{3}	5.0×10^{4}
Thyroid:apical membrane transport						
K_m (ng/L)	1.0×10^{8}	1.0×10^{8}	1.0×10^{8}	1.0×10^{8}	1.0×10^{8}	1.0×10^{8}

$V_{max}C$ (ng/hr per kg)	2.0×10^{4}	1.0×10^{4}	1.0×10^{4}	1.0×10^{4}	1.0×10^{4}	2.5×10^{5}
Gastric: NIS active transport						
K_m (ng/L)	1.7×10^{5} [b]	1.0×10^{5}	1.0×10^{5}	1.5×10^{5}	—	2.0×10^{5} [b]
$V_{max}C$ (ng/hr per kg)	2.0×10^{4} [b]	8.0×10^{5}	1.0×10^{5}	1.0×10^{6}	1.0×10^{6}	1.0×10^{5} [b]
	Perchlorate-Specific Parameters					
Skin:NIS active transport						
K_m (ng/L)	1.8×10^{5}	1.0×10^{5}	1.0×10^{5}	1.5×10^{5}	1.5×10^{5}	2.0×10^{5}
$V_{max}C$ (ng/hr per kg)	5.0×10^{5}	6.0×10^{5}	4.0×10^{5}	8.0×10^{5}	8.0×10^{5}	1.0×10^{6}
Mammary: active transport						
K_m (ng/L)	—	1.0×10^{5}	—	1.5×10^{5}	—	—
$V_{max}C$ (ng/hr per kg)	—	2.2×10^{4}	—	2.0×10^{4}	—	—
Milk: active transport						
K_m (ng/L)	—	—	—	1.0×10^{6}	—	—
$V_{max}C$ (ng/hr per kg)	—	—	—	2.0×10^{4}	—	—
Placenta active transport						
K_m (ng/L)	—	1.0×10^{5}	—	—	—	—
$V_{max}C$ (ng/hr per kg)	—	6.0×10^{4}	—	—	—	—
Transfer placenta to fetus (L/hr per kg)	—	0.065	—	—	—	—
Transfer fetus to placenta (L/hr per kg)	—	0.12	—	—	—	—
Plasma-protein binding						
K_m (ng/L)	1.1×10^{4}	1.0×10^{4}	1.5×10^{4}	1.0×10^{4}	1.0×10^{4}	1.8×10^{4}
$V_{max}C$ (ng/hr per kg)	3.4×10^{3}	4.0×10^{3}	1.5×10^{3}	9.0×10^{3}	2.0×10^{3}	5.0×10^{2}
Dissociation constant (L/hr per kg)	0.032	0.034	0.01	0.034	0.01	0.025
Urinary clearance (L/hr per kg)	0.07	0.07	—	0.07	0.0075	0.1265

TABLE E-1 *(Continued)*

Parameter	Male Rat	Pregnancy (Rat, GD 0-21)[a]		Lactation (Rat, PND 0-18)		Adult
		Dam	Fetus	Dam	Neonate	Human
Fraction pup urine ingested by dam	—	—	—	0.80	—	—
Iodide-Specific Parameters						
Partition coefficients (unitless)						
Poorly perfused:plasma	0.21	0.21	0.21	0.21	0.21	0.21
Richly perfused:plasma	0.40	0.40	0.40	0.40	0.40	0.40
Fat:plasma	0.05	0.05	—	0.05	0.05	0.05
Kidney:plasma	1.00	1.09	1.09	1.09	1.09	1.09
Liver:plasma	0.44	0.44	0.44	0.44	0.44	0.44
Gastric tissue:gastric blood	1.00[b]	1.00	1.00	1.00	1.20	0.50[b]
Gastric contents:gastric tissue	3.50[b]	2.00	2.00	1.00	1.00	3.50[b]
Skin tissue:skin blood	0.70	0.70	0.70	0.70	1.00	0.70
Red blood cells:plasma	1.00	1.00	1.00	1.00	1.00	1.00
Placenta:plasma	—	0.40	—	—	—	—
Mammary gland:plasma	—	0.66	—	—	—	—
Mammary tissue:mammary blood	—	—	—	0.80	—	—
Mammary tissue:milk	—	—	—	1.00	—	—
Thyroid follicle:stroma	0.15	0.15	0.15	0.15	0.15	0.15
Thyroid lumen:follicle	8.00	7.00	7.00	7.00	7.00	7.00
Permeability area cross products (L/hr per kg)						
Gastric blood-gastric tissue	1.00[b]	0.80	0.10	0.80	0.04	0.20[b]
Gastric tissue-gastric contents	0.10[b]	0.60	0.30	0.60	0.09	2.00[b]

Skin blood-skin tissue	0.10	0.10	0.02	0.20	0.02	0.06
Plasma-red blood cells	1.00	1.00	1.00	1.00	1.00	1.00
Placenta blood-placenta tissue	—	0.005	—	—	—	—
Mammary blood-mammary tissue	—	0.01	—	0.02	—	—
Mammary tissue-milk	—	—	—	0.02	—	—
Thyroid follicle-stroma	1.0×10^{-4}	1.0×10^{-4}	1.0×10^{-4}	1.0×10^{-4}	1.0×10^{-4}	1.0×10^{-4}
Thyroid lumen-follicle	4.0×10^{-7}	4.0×10^{-7}	4.0×10^{-4}	1.0×10^{-4}	1.0×10^{-4}	1.0×10^{-4}
Thyroid: NIS active transport						
K_m (ng/L)	4.0×10^{6}	4.0×10^{6}	4.0×10^{6}	4.0×10^{6}	4.0×10^{6}	4.0×10^{6}
$V_{max}C$ (ng/hr per kg)	5.4×10^{6}	4.4×10^{4}	$0\text{-}5.0 \times 10^{4}$	5.0×10^{4}	1.3×10^{4}	1.5×10^{5}
Thyroid: apical membrane transport						
K_m (ng/L)	1.0×10^{9}	1.0×10^{9}	1.0×10^{9}	1.0×10^{9}	1.0×10^{9}	1.0×10^{9}
$V_{max}C$ (ng/hr per kg)	4.0×10^{6}	4.0×10^{6}	4.0×10^{6}	6.0×10^{7}	6.0×10^{7}	1.0×10^{8}
Thyroid:hormone production (L/hr per kg)	0.1	0.03	—	0.1	0.06	—
Thyroid: hormone secretion (L/hr per kg)	1.2×10^{6}	1.0×10^{-6}	—	7.0×10^{-7}	1.0×10^{-6}	—
Deiodination of thyroid hormone (L/hr per kg)	—	—	—	0.02	0.025	—
Gastric:NIS active transport						
K_m (ng/L)	$4.0 \times 10^{6\,b}$	$4.0 \times 10^{6\,b}$	$4.0 \times 10^{6\,b}$	$4.0 \times 10^{6\,b}$	—	$4.0 \times 10^{6\,b}$
$V_{max}C$ (ng/hr per kg)	$2.0 \times 10^{6\,b}$	1.0×10^{6}	2.0×10^{6}	2.0×10^{6}	2.0×10^{6}	$9.0 \times 10^{5\,b}$
Skin: NIS active transport						
K_m (ng/L)	4.0×10^{6}	4.0×10^{6}	4.0×10^{6}	4.0×10^{6}	4.0×10^{6}	4.0×10^{6}
$V_{max}C$ (ng/hr per kg)	5.0×10^{5}	6.0×10^{4}	7.0×10^{5}	4.0×10^{5}	2.5×10^{5}	7.0×10^{5}
Mammary: active transport						
K_m (ng/L)	—	4.0×10^{6}	—	4.0×10^{6}	—	—
$V_{max}C$ (ng/hr per kg)	—	4.0×10^{4}	—	8.0×10^{5}	—	—

TABLE E-1 *(Continued)*

Parameter	Male Rat	Pregnancy (Rat, GD 0-21)[a] Dam	Pregnancy (Rat, GD 0-21)[a] Fetus	Lactation (Rat, PND 0-18) Dam	Lactation (Rat, PND 0-18) Neonate	Adult Human
Milk: active transport						
K_m (ng/L)	—	—	—	1.0×10^7	—	—
$V_{max}C$ (ng/hr per kg)	—	—	—	4.0×10^5	—	—
Placenta active transport						
K_m (ng/L)	—	4.0×10^6	—	—	—	—
$V_{max}C$ (ng/hr per kg)	—	5.5×10^4	—	—	—	—
Transfer placenta to fetus (L/hr per kg)	—	0.06	—	—	—	—
Transfer fetus to placenta (L/hr per kg)	—	0.12	—	—	—	—
Plasma protein binding						
K_m (ng/L)	—	—	—	1.0×10^5	1.0×10^5	—
$V_{max}C$ (ng/hr per kg)	1.0×10^2 [e]	—	—	1.5×10^3	5×10^2	—
Dissociation constant (L/hr per kg)	—	—	—	0.09	0.05	—
Urinary clearance (L/hr per kg)	0.05	0.03	—	0.06	0.012	0.1
Fraction pup urine ingested by dam	—	—	—	0.8	—	—

[a]Pregnancy models covered both embryonic and fetal development up to gestation day 21; tissue samples were taken only from developing fetuses (gestation day 20-21) in pharmacokinetic studies. Earliest day for fetal tissue volume data is gestation day 11, so exponential curves were used to simulate tissue growth during embryonic development. Final equations for fetal growth were adjusted by measured pup body weights at birth (or when samples were collected from kinetic studies).

[b]Male rat and adult human GI refers only to stomach.

[c]Data from Williams and Leggett (1989).

[d]$V_{max}C$ is V_{max} scaled by body weight according to equation V_{max} (mg/hr) = [$V_{max}C$ (mg/hr per kg)][body weight (kg)$^{0.70}$].

[e]It was unclear why value for capacity of plasma-protein binding of iodide (V_{max}) was reported in adult male rat model but protein binding was included only in lactating rat PBPK model.

Abbreviations: BW, body weight; F, female; GD, gestation day; K_m, Michaelis-Menten constant; hr, hour; kg, kilogram; L, liter; M, male; NIS, sodium-iodide symporter; ng, nanogram; PND, postnatal day; VGI, volume of gastrointestinal tract; V_{max}, maximum capacity for binding or transport; $V_{max}C$, maximum capacity for binding or transport scaled by body weight; VSk, volume of skin.

containing tissue were not substantially different between the various rat and human PBPK models summarized in Table E-1. However, when attempting to simulate the radioiodide-uptake data of Greer et al. (2000, 2002), Merrill (2001b) determined that a nearly 10-fold range in NIS capacity existed between subjects, which is not unusual for many biochemical processes. In fact, the baseline thyroid radioiodide uptake varied by a factor of 3-5 among the volunteers. In addition to potential differences in NIS protein expression between subjects, there are potential environmental influences on the variability of the kinetic data, including differences in endogenous iodide concentrations resulting from differences in diets, the timing of blood collections relative to meals, the presence of other potential inhibitors of NIS, and other factors that were not controlled for. As discussed in Chapter 3 and Appendix D, the differences in absolute capacities of the NIS or differences in basal iodide concentrations between humans *by themselves* should not have an important effect on intersubject sensitivity to the rates of perchlorate inhibition of iodide uptake. Regardless, variations in thyroid parameter values have little effect on serum concentrations of perchlorate (and iodide) that EPA used in deriving the RfD, because the thyroid is small.

Plasma-protein binding is often a source of substantial species differences in chemical disposition, as was evident for perchlorate. On the basis of the serum data of Greer et al. (2000), humans had a lower capacity for binding perchlorate than did rats (for example, the capacity for plasma-protein binding was 500 ng/hr per kilogram in humans vs 3,400 ng/hr per kilogram in rats), whereas the binding affinities and dissociation constants were similar. Likewise, urinary clearance may reflect substantial species differences; in the human PBPK model, the urinary clearances of both anions were about twice as high as in the rat, possibly partially because of decreased plasma-protein binding, at least for perchlorate.

The chemical-specific parameters were either scaled from the rat or developed from the iodide kinetic data of Hays and Solomon (1965) and the perchlorate- and iodide-uptake data and inhibition measurements of Greer et al. (2000, 2002). Model validations (simulations of data from independent experiments that were not used in model development) were based on simulations of the urinary clearance of perchlorate in healthy males after oral doses of 9.07-20 mg/kg (Eichen 1929; Durand 1938; Kamm and Drescher 1973), serum perchlorate concentrations after drinking-water exposures at 12 mg/kg per day (unpublished results provided to E. Merrill by Dr. Brabant, Hanover, Germany [Merrill 2001b]), and iodide uptake in the thyroid of a male with Graves disease when the maximum capacity of the NIS was increased about 10 times that in healthy subjects (Stanbury and Wyngaarden 1952). However, the model underpredicted the degree of inhibition of thyroid radioiodide uptake after dosing with 100 mg of potassium perchlorate; this suggests that the increased inhibition of iodide uptake in Graves disease may not be simply due to the affinity or capacity of the NIS (no perchlorate or iodide kinetics in serum or urine were evaluated in this study).

Overall, the adult male human PBPK model provided a reasonable description of the available human perchlorate and iodide kinetics data over doses spanning several orders of magnitude (0.02-12 mg/kg). As discussed above for the male rat, the human skin compartment may be important in controlling the kinetics of both anions, although to a much smaller extent than in rats because of its smaller fraction of total body weight. Regardless, including a skin compartment was critical to the human model simulations of serum perchlorate concentrations, so the paucity of human data on the disposition of perchlorate and iodide in skin remains a concern for future research. Although no formal sensitivity analysis was performed on the human PBPK model, it is likely that, in addition to the skin compartment, urinary clearance of both anions and the plasma-protein binding of perchlorate may be important for additional future research. Furthermore, the PBPK model was developed for adult males and females (primarily healthy subjects although one subject with Graves disease was simulated) but not for pregnant or lactating females, human fetuses, neonates, or children. To assist in validating the parallelogram approach used to derive human equivalent exposures, discussed in Chapter 4, consideration should be given to refining the iodide PBPK model to incorporate data from biomonitoring studies, such as Soldin et al. (2003) and Hollowell et al. (1998), that include the analysis of iodide in pregnant women.

PBPK MODEL FOR PREGNANT RATS AND FETUSES

During gestation and early infancy, thyroid hormones are needed for normal development (see Chapter 2). Although the critical periods for effective disruption due to the inhibition of thyroid iodide uptake by perchlorate are not known, it is likely that the developing fetus becomes directly susceptible to perchlorate when the thyroid begins to sequester iodide and secrete hormones. That occurs by 12 weeks of gestation in humans or around 17-20 days in rats (Clewell et al. 2003a). The fetus is potentially indirectly susceptible earlier if the mother's hormone production is compromised. Because perchlorate can competitively inhibit iodide uptake in the maternal thyroid, placental iodide transfer, or fetal thyroid uptake, Clewell et al. (2003a) developed a PBPK model covering the full gestation period in rats.

Clewell et al. (2003a) extended the male rat PBPK model to the pregnant female rat by using pregnant female rat-specific physiology over gestation days 2-20 and adding placental and mammary tissue compartments. The maternal model was linked with a fetal model that was similar in structure to the adult model (without the fat, placenta, and mammary tissues), as shown in Figure E-5. To simplify the model, all fetuses from a single litter were lumped together in the model structure (for example, each tissue compartment was multiplied by the number of fetuses in each experiment to simulate the available data). That simplification was necessary to simulate the kinetic data in fetal samples that were too small and had to be pooled across a litter for analyses. Although a kidney compartment was included in the fetal model, urinary excretion was not included, because urine production is not well developed until after birth.

Pregnancy is a remarkably dynamic process in which changes occurr rapidly in both the pregnant animal and its developing offspring. Those changes have the potential for affecting the delivery of perchlorate (and iodide) to its target site at the appropriate time for effects to occur. In the PBPK models just described for adult rats and humans, the volume of each tissue and its corresponding blood flow were treated as constants. However, during pregnancy, some tissue volumes and blood flow rates change in the mother and the embryo and fetus and must be treated as variables rather than constants in the PBPK model. As a result, Clewell et al. (2003a) based their physiologic model on the work of O'Flaherty et al. (1992) and Fisher et al. (1989), who derived detailed equations for each tissue's unique and dynamically changing growth rates and blood perfusion rates over the entire gestation period in rats. Those models have been widely used as the structural basis of most rat pregnancy PBPK models developed over the last 10 years (Corley et al. 2003).

FIGURE E-5 Diagram of PBPK model for perchlorate and iodide distribution in pregnant rat and developing fetus (adapted from Clewell et al. [2003a]). Bold arrows indicate active transport of perchlorate and iodide by NIS in thyroid, skin, gastric mucosa, mammary gland, and placenta and by apical iodide channels in thyroid. Model compartments that were added to adult rat PBPK model of Merrill et al. (2003) are designated by shading. Abbreviations: ClO_4^-, perchlorate ion; GI, gastrointestinal; IV, intravenous; NIS, sodium-iodide symporter; PBPK, physiologically based pharmacokinetic; RBCs, red blood cells.

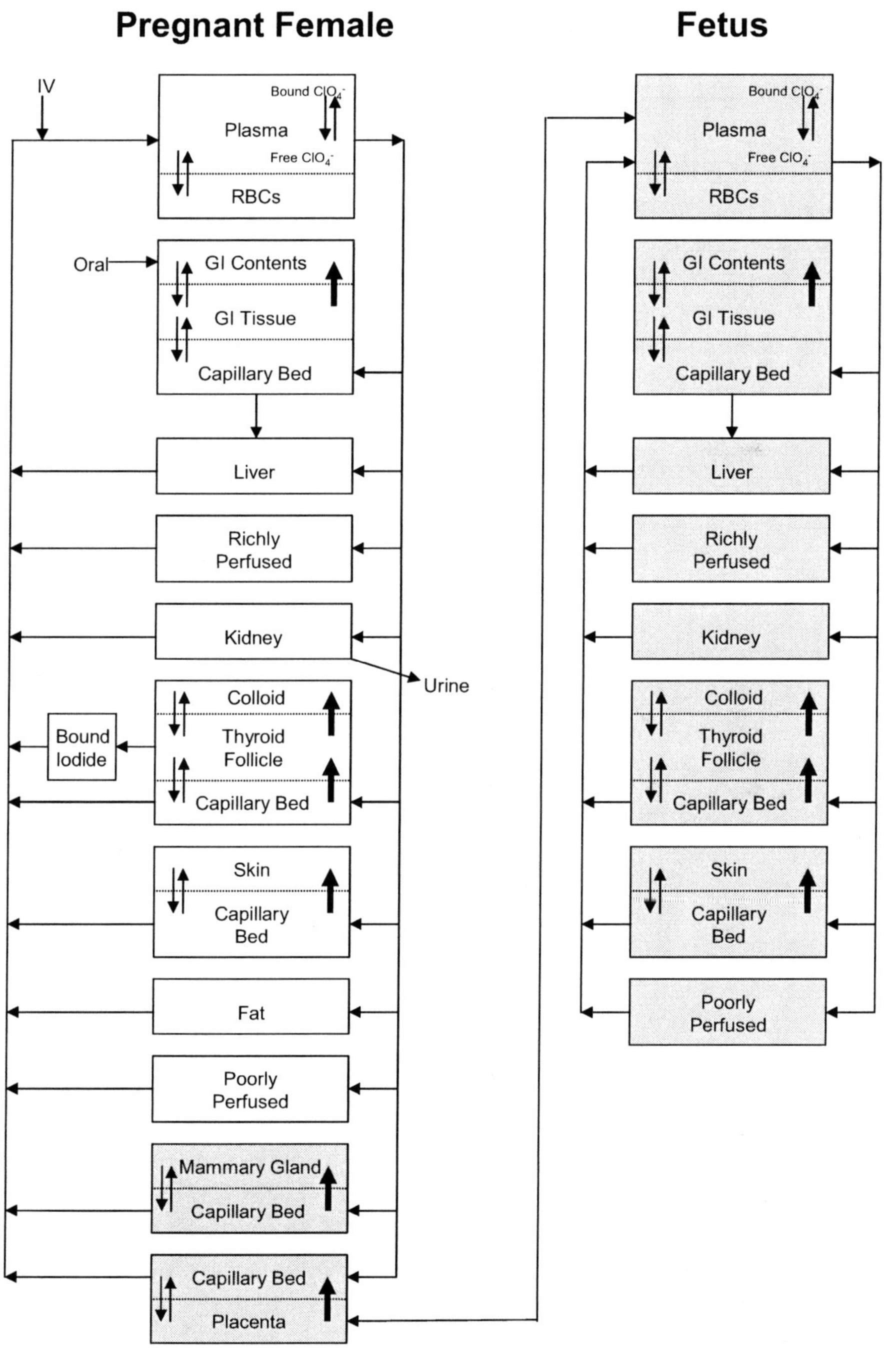
Pregnant Female
Fetus
IV
Oral
Bound ClO$_4^-$
Plasma
Free ClO$_4^-$
RBCs
GI Contents
GI Tissue
Capillary Bed
Liver
Richly Perfused
Kidney
Urine
Bound Iodide
Colloid
Thyroid Follicle
Capillary Bed
Skin
Capillary Bed
Fat
Poorly Perfused
Mammary Gland
Capillary Bed
Capillary Bed
Placenta
Bound ClO$_4^-$
Plasma
Free ClO$_4^-$
RBCs
GI Contents
GI Tissue
Capillary Bed
Liver
Richly Perfused
Kidney
Colloid
Thyroid Follicle
Capillary Bed
Skin
Capillary Bed
Poorly Perfused

As with the adult male rat model of Merrill (2001b), up-regulation of maternal thyroid NIS was included as a heuristic approximation of feedback control by the HPT axis because of the inhibition of thyroid iodide uptake by perchlorate (that is, no attempts were made to describe the time-dependent up-regulation or other physiologic changes associated with it). Although fetal thyroids begin to secrete hormone during gestations days 17-20, organification of thyroid iodide was not included in the fetal thyroid model, as it is in the mother, because of a lack of quantitative data on production and secretion rates.

The mammary gland, which also has NIS, was not shown to concentrate perchlorate and iodide over peak serum concentrations during gestation although the concentrations of both anions were observed by Clewell et al. (2003a) to remain higher in mammary tissue than in serum during the clearance phase. Therefore, the mammary gland was divided into two compartments: mammary blood and mammary tissue with both NIS transport and passive-diffusion equations included, as for the skin compartment. Likewise, the placenta, which also contains the NIS, was modeled as two compartments.

Transport of nutrients and xenobiotics to the embryo and fetus occurs across both the yolk sac and chorioallantoic placentas, which vary substantially among species in function and development. For example, iron and proteins are transported across the yolk sac in rats and across the chorioallantoic placenta in humans. Thus, several PBPK models that describe the transfer of xenobiotics between the mother and the developing embryo and fetus include both types of placenta, which develop at different rates. Because of a lack of appropriate data, Clewell et al. (2003a) combined both types of placenta into one placental compartment with empirical first-order rate constants governing the transfer of perchlorate and iodide between the placenta and the fetal plasma compartments (see Figure E-5). Future research could include a more realistic understanding of the placental transfer of perchlorate and iodide.

To determine values of several chemical-specific parameters for their model, Clewell et al. (2003a) conducted a series of experiments, including drinking-water kinetic studies with perchlorate in pregnant Sprague-Dawley rats over gestation days 2-20, radioiodide kinetic studies after IV administrations on gestation day 20, and radioiodide-inhibition studies on gestation day 20. Those studies were used to develop and refine estimates of NIS transport parameter values, thyroid apical membrane transport parameter values, partition coefficients, permeability-area cross products for diffusion, plasma-protein binding (perchlorate only), and urinary clearance. For several tissues, there was a biologic basis for expecting differences in perchlorate and iodide kinetic parameter values in male versus pregnant female rats (such as plasma-protein binding, thyroid transport, and skin transport). For other tissues, model parameter values were the same (or nearly so) between adult male and pregnant female rats. The committee does not see such differences as detracting from the utility of the PBPK models for simulating elements important to the disposition of perchlorate and iodide as a function of life stage.

The predictions of the PBPK model were validated with datasets that were not used to develop the model. Several published datasets describing the kinetics of iodide in pregnant rats were reasonably well described by the PBPK model, especially given the rapidly changing dynamics of gestation. The ability of the model to predict the perchlorate-induced inhibition of thyroid iodide uptake in published studies and those conducted by Clewell et al. (2003a) requires accurate simulations of both perchlorate and iodide kinetics. Thus, those datasets served as an additional validation of the perchlorate and iodide PBPK models.

Clewell et al. (2003a) also performed a sensitivity analysis to determine which model parameters had the greatest effect on simulations of serum perchlorate concentrations and thyroid iodide uptake. As with the adult male rat PBPK model, maternal serum perchlorate is sensitive to serum-protein binding and urinary-clearance parameters. Fetal serum simulations were sensitive to plasma transfer rates, placental NIS, placental diffusion, fetal serum-protein binding, and maternal urinary-clearance parameters. As discussed above for the male rat, parameters identified by the sensitivity analysis present important research opportunities. The inhibition of thyroid iodide uptake was sensitive to many more parameters. The authors appropriately concluded that that was because thyroid iodide uptake simulations were for a specific point in time and were thus more sensitive to the changing kinetics of *bth* perchlorate and iodide.

PBPK MODEL FOR LACTATING RATS AND NEONATES

Neonatal hormone feedback control is independent of the dam, and the thyroid gland continues to develop after birth in the rat. But, the dam still controls both perchlorate and iodide transfer to the offspring via nursing. Therefore, Clewell et al. (2003b) extended their rat gestation PBPK model to include lactation transfer over postnatal days 0-10. As with gestation models, lactation-transfer PBPK models must contend with rapid changes in both maternal and neonatal physiology. There are fewer PBPK models that cover this period of development than there are gestation models

(Corley et al. 2003). Therefore, Clewell et al. (2003b) relied on the initial framework described by Shelley et al. (1988) and the earliest model that incorporated physiologic changes of the rat dam and neonate by Fisher et al. (1990) as the structural and physiologic basis of the perchlorate and iodide PBPK models. Additional compartments that were necessary to describe perchlorate and iodide kinetics and interactions were either taken from the adult rat and pregnant rat PBPK models described above or modified to specifically describe the data available on lactation transfer of perchlorate and iodide. The latter modifications are described below.

During gestation, the NIS in the mammary gland delays the clearance of iodide (and perchlorate) from mammary tissues although these anions typically do not rise above peak serum concentrations. During lactation, however, iodide is concentrated in the milk. Inhibition of iodide uptake by perchlorate not only decreases the iodide available in milk but also transfers perchlorate, which can inhibit the uptake of iodide in the thyroid of the nursing pup (Yu et al. 2001). Thus, the two-compartment model for the mammary gland used in the pregnancy model of Clewell et al. (2003a) was expanded to three compartments: mammary blood, mammary tissue, and milk at parturition (see Figure E-6). Similar to NIS activity in the thyroid, NIS transport of iodide and perchlorate occurs between mammary blood and mammary tissue. A second mechanism reportedly exists to transport iodide against a concentration gradient into milk, so Clewell et al. (2003b) incorporated equations for the competitive transport of perchlorate and iodide (designated by bold arrows in Figure E-6) between compartments with passive diffusion that is driven by concentration gradients between compartments. Perchlorate and iodide are then transferred from the milk to the GI contents of the pup with a first-order rate of transfer, and both anions are returned to the dam (GI contents) via pup urine with a first-order transfer rate to simulate grooming by the dam.

The remaining maternal tissue compartments, except plasma, were structured similarly to the way they were in the pregnancy model with changing physiology over postnatal days 0-10 based on the lactation model of Fisher et al. (1990). The neonatal model was similar to the maternal model except for the absence of mammary tissues. In contrast with the developing fetus, a fat compartment was needed because this tissue begins to grow after birth. To simplify the model, all pups from a given litter were combined in the model structure (that is, total body weight was multiplied by the number of pups in each litter).

Thyroid secretion of hormones was modeled in a fashion similar to that used in the adult rat and pregnancy models with one major exception. In the lactation model, Clewell et al. (2003b) added deiodination of thyroid hormones to the plasma compartment as a surrogate of tissue deiodination and included competitive protein binding between iodide and perchlorate in plasma (Figure E-7). It is not clear why that is needed in the lactation model but not in the adult rat, adult human, and pregnant rat models.

To develop the lactation-transfer PBPK model, Clewell et al. (2003b) conducted perchlorate drinking-water exposure studies in Sprague-Dawley rats covering the period from gestation day 2 through postnatal day 5 or 10 and radioiodide kinetic and inhibition studies in lactating rats and neonates on postnatal day 10. Kinetic parameter values for the perchlorate model were estimated from the maternal drinking-water study for both dams and neonates. For the iodide model, chemical-specific parameter values were obtained for the dams and neonates that were each dosed independently. Chemical-specific model parameter values (Table E-1) in tissues other than the mammary gland and blood compartment were kept as similar as possible to those of the adult rat and pregnant rat models although some were adjusted to fit the perchlorate and iodide kinetic studies conducted by Clewell et al. (2003b). Up-regulation of thyroid NIS to simulate the drinking-water kinetic studies was performed as described for the adult male rat PBPK model.

Model simulations of neonatal exposure via nursing were assumed to be continuous; in gavage experiments, the dose was introduced as a bolus directly into GI contents. Drinking-water exposure for the perchlorate studies was modeled as 12-hr/day exposure to reflect nocturnal feeding behaviors (6 p.m. to 6 a.m.); simulations of dietary iodide studies, which were reported only for lactating rats and their pups, were based on the assumption of 24-hr/day intake. The reasons for the differences in maternal exposure (12 hr/day vs 24 hr/day) were not explained by the authors. Thus, the radioiodide, dietary iodide, and perchlorate models were operated independently with interactions between the three models occurring at the level of the NIS and, although not explicitly stated by the authors, presumably via competitive plasma-protein binding.

The resulting models were able to simulate maternal and neonatal perchlorate and iodide kinetics whether only the dams were exposed or the neonates were exposed directly. Validation of the model simulations was based on several datasets from the literature that were not used to estimate model parameter values and on the thyroid iodide uptake inhibition studies. The lactation-transfer model was able to simulate perchlorate and iodide kinetics and inhibition of NIS transport in the thyroid reliably over a nearly complete period of lactation (up to postnatal day 20).

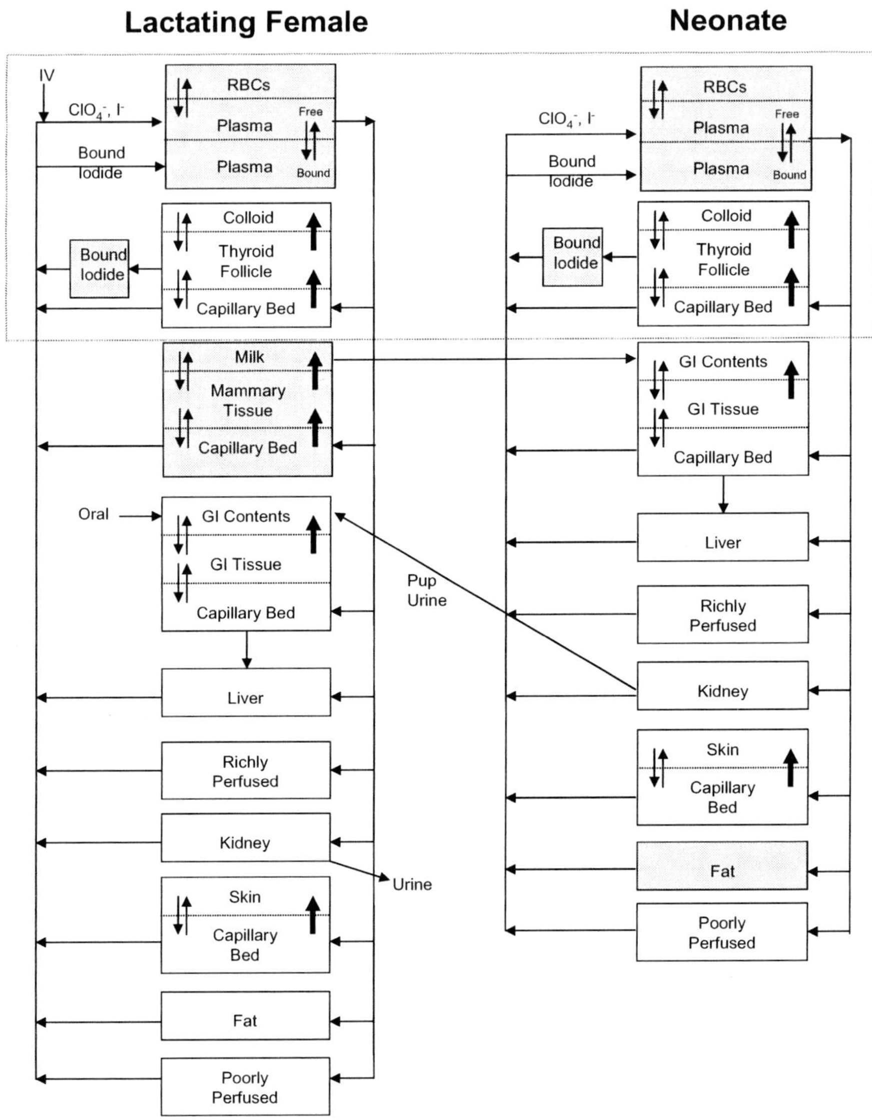

FIGURE E-6 Diagram of PBPK model for perchlorate and iodide in lactating rat and neonate (adapted from Clewell et al. [2003b]). Bold arrows indicate active transport of perchlorate and iodide NIS in thyroid, skin, gastric mucosa, mammary gland, and placenta and by apical iodide channels in thyroid and mammary gland. Model compartments that were altered or added to pregnancy model of Clewell et al. (2003a) are designated by shading. See Figure E-7 for more detailed diagram

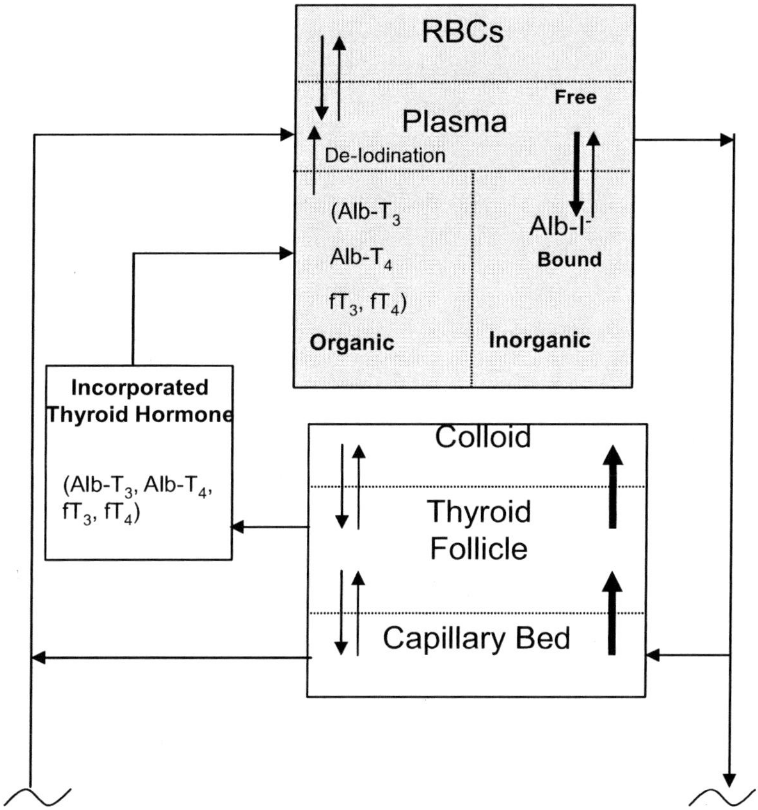

FIGURE E-7 Diagram of thyroid and blood subcompartments from iodide PBPK model of lactating and neonatal rat (adapted from Clewell et al. 2003b). Bold arrows indicate active transport of iodide. Model subcompartments that were altered or added to pregnancy model of Clewell et al. (2003a) are designated by shading. Thyroid hormones (free and bound) were lumped together and secreted by thyroid follicle into plasma where first-order deiodination reaction (a surrogate of tissue metabolism) releases inorganic iodide, which can compete with perchlorate for binding to plasma proteins. Abbreviations: Alb, albumin; I^-, iodide; fT_3, free triiodothyronine; fT_4, free thyroxine; T_3, triiodothyronine; T_4, thyroxine.

---

of thyroid-blood compartments of iodide model outlined in this figure. Abbreviations: ClO_4^-, perchlorate ion; GI, gastrointestinal; I^-, iodide; IV, intravenous; NIS, sodium-iodide symporter; PBPK, physiologically based pharmacokinetic; RBCs, red blood cells.

A rudimentary dietary and drinking-water exposure iodide PBPK model was developed to simulate "endogenous" iodide from the radioidide studies in the lactating rat and nursing pups, so the authors conducted additional simulations to determine the effect of changing dietary iodide on predicted acute radioiodide kinetics and perchlorate-induced inhibition of dietary uptake. Their simulations were consistent with the committee's calculations for humans discussed in Chapter 3 and Appendix D. Changes in dietary iodide, even at concentrations over at least 100 times those used in standard laboratory diets, should have no effect on the sensitivity of the thyroid to the inhibition of iodide uptake by perchlorate in the lactating rat.

As with the other PBPK models, Clewell et al. (2003b) also conducted a sensitivity analysis to determine which model parameters had the greatest effects on the predictions of serum perchlorate and thyroid iodide uptake inhibition. The results were similar to those of previous model sensitivity analyses and included parameters associated with plasma-protein binding, renal clearance, and rates of transfer between dams and nursing pups.

COMPARISONS OF INTERNAL-DOSE SURROGATES ACROSS SPECIES AND LIFE STAGE

Clewell et al. (2003b) compared the results of simulating the serum concentrations of perchlorate measured as area under the curve after drinking-water exposure with results of simulating the inhibition of thyroid iodide uptake after acute exposure to perchlorate in each of the PBPK models described above for the rat. The results of area under the curve simulations for serum perchlorate are shown in Figure E-8, and those of simulated inhibition of thyroid iodide uptake in Figure E-9.

Clewell et al. (2003b) did not report whether up-regulated NIS parameter values were used in the simulations. Regardless, on the basis solely of the two internal-dose metrics, the lactating dam may be at greatest risk from perchlorate exposure if serum perchlorate area under the curve is used as the internal-dose surrogate, whereas the fetus appears to be the most sensitive if inhibition of thyroid iodide uptake is used as the surrogate. Clewell et al. (2003b) attributed the increased sensitivity of the lactating dam to the increase in protein binding of perchlorate in plasma (see Table E-1); fetal sensitivity to iodide uptake inhibition is multifactorial and is influenced by maternal transfer of perchlorate, inhibition of iodide transfer at the placenta, and other factors. EPA (2002b) used such simulations to calculate human equivalent exposures that corresponded to the internal-dose surrogates.

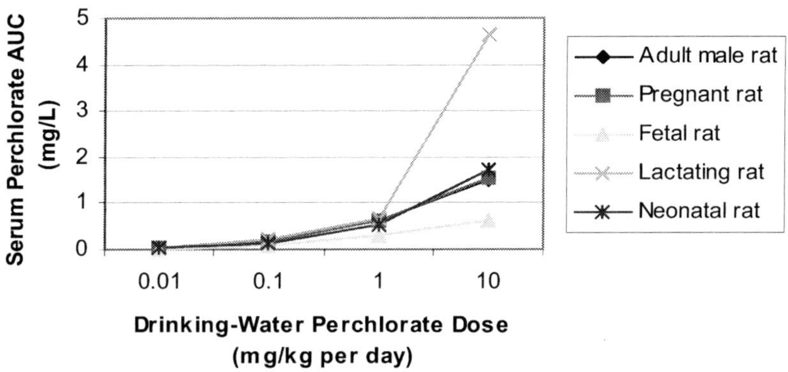

FIGURE E-8 Dose-response simulations of area under the curve for concentrations of perchlorate in serum of adult male, pregnant, fetal, lactating, and neonatal rats after drinking-water exposure as described in Clewell et al. (2003b). Abbreviations: AUC, area under the curve; kg, kilogram; mg, milligram; mg/L, milligrams per liter; mg/kg/day, milligrams per kilogram of body weight per day.

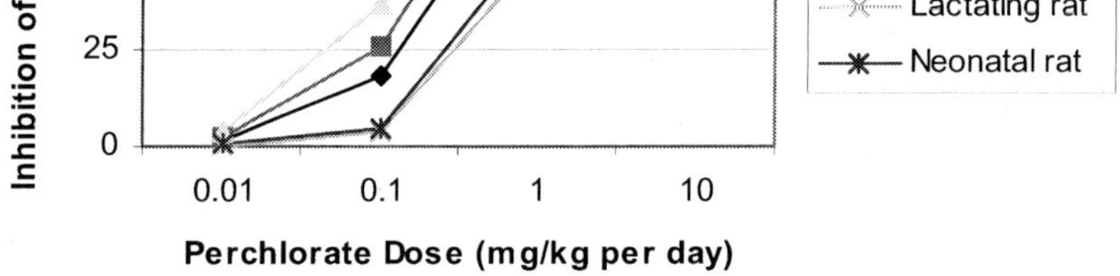

FIGURE E-9 Dose-response simulations of inhibition of thyroid iodide uptake in adult male, pregnant, fetal, lactating, and neonatal rats after acute perchlorate exposure as described in Clewell et al. (2003b). Abbreviation: mg/kg/day, milligram per kilogram of body weight per day.

SUMMARY AND CONCLUSIONS

Although the PBPK models simplified the interactions between perchlorate and iodide to a series of rate-limiting steps, their underlying biologic basis effectively reduced the uncertainties associated with extrapolating blood concentrations of perchlorate and inhibition of thyroid iodide uptake across species, dose, and routes of exposure for adult animals. Furthermore, the parallelogram approach discussed in Chapter 4, applied where models have been validated in the adult, pregnant, and lactating rat, and adult humans, serves as a useful tool for constraining extrapolations to adult human females during pregnancy or lactation. However, given the important species differences in developmental biology and the current inability to validate extrapolation to human fetuses and neonates, such an approach should be used with caution for these potentially sensitive populations. The committee concludes that it was appropriate for both DOD and EPA to limit animal-to-human extrapolations for pregnancy and lactation to maternal serum perchlorate concentrations or interactions with iodide in maternal thyroid NIS.

The suite of PBPK models developed by DOD for the adult rat, adult human, pregnant rat and fetus, and lactating rat and neonate represent the current state-of-the-science approach for integrating available animal (rat) and human data on the disposition of perchlorate and iodide and their interactions at the level of the thyroid NIS. Although many of the PBPK model parameter values had to be estimated from a limited set of in vivo pharmacokinetic studies rather than independently measured, enough studies were available for validation of model simulations over a broad range of perchlorate doses and iodide concentrations to lend confidence to the application of the models for extrapolating internal-dose surrogates from animals to adult humans. If future studies are conducted to elucidate further the toxicity or mode of action of perchlorate in animal models, consideration should be given to updating the PBPK models because they provide a convenient framework to assemble current knowledge on the disposition of perchlorate in the body and on how it may interact with iodide at various stages of development.

REFERENCES

Brown, R.P., M.D. Delp, S.L. Lindstedt, L.R. Rhomberg, and R.P. Beliles. 1997. Physiological parameter values for physiologically based pharmacokinetic models. Toxicol. Ind. Health 13(4):407-484.

Clewell, R.A. 2001a. Consultative letter AFRL-HE-WP-CL-2001-0006. Physiologically Based Pharmacokinetic Model for the Kinetics of Perchlorate-Induced Inhibition of Iodide in the Pregnant Rat and Fetus. Memorandum to Annie M. Jarabek, NCEA, U.S. Environmental Protection Agency, Research Triangle Park, NC, from Rebecca Clewell, Air Force Research Laboratory, Department of the Air Force, Wright-Patterson Air Force Base, OH. May 10, 2001.

Clewell, R.A. 2001b. Consultative letter, AFRL-HE-WP-CL-2001-0007. Physiologically-Based Pharmacokinetic Model for the Kinetics of Perchlorate-Induced Inhibition of Iodide in the Lactating and Neonatal Rat. Memorandum to Annie M. Jarabek, NCEA, U.S. Environmental Protection Agency, Research Triangle Park, NC, from Rebecca Clewell, Air Force Research Laboratory, Department of the Air Force, Wright-Patterson Air Force Base, OH. May 24, 2001.

Clewell, R.A., E.A. Merrill, P.J. Robinson. 2001. The use of physiologically-based models to integrate diverse data sets and reduce uncertainty in the prediction of perchlorate and iodide kinetics across life stages and species. Toxicol. Ind. Health 17(5-10):210-222.

Clewell, R.A., E.A. Merrill, K.O. Yu, D.A. Mahle, T.R. Sterner, D.R. Mattie, P.J. Robinson, J.W. Fisher, and J.M. Gearhart. 2003a. Predicting fetal perchlorate dose and inhibition of iodide kinetics during gestation: a physiologically-based pharmacokinetic analysis of perchlorate and iodide kinetics in the rat. Toxicol. Sci. 73(2):235-255.

Clewell, R.A., E.A. Merrill, K.O. Yu, D.A. Mahle, T.R. Sterner, J.W. Fisher, and J.M. Gearhart. 2003b. Predicting neonatal perchlorate dose and inhibition of iodide uptake in the rat during lactation using physiologically based pharmacokinetic modeling. Toxicol. Sci. 74(2):416-436.

Clewell, R.A., E.A. Merrill, L. Narayanan, J.M. Gearhart, and P.J. Robinson. 2004. Evidence for competitive inhibition of iodide uptake by perchlorate and translocation of perchlorate into the thyroid. Int. J. Toxicol. 23(1):17-23.

Corley, R.A., T.J. Mast, E.W. Carney, J.M. Rogers, and G.P. Daston. 2003. Evaluation of physiologically-based modeling of pregnancy and lactation for their application in children's health risk assessments. Crit. Rev. Toxicol. 33(2):137-211.

Durand, J. 1938. Recherches sur l'elimination des perchlorates, sur leur repartition dans les organs et sur leur toxicite [Research on the elimination of perchlorate, its distribution in organs and its toxicity]. Bull. Soc. Chim. Biol. 20:423-433 (as cited in Stanbury and Wyngarrden 1952).

Eichler, O. 1929. Zur Pharmakologie der Perchloratwirkung [The pharmacology of the perchlorate effect]. Naunyn-Schmeideberg's Arch. Exp. Path. U. Pharmak. 144:251-260 (as cited in Stanbury and Wyngaarden 1952).

EPA (U.S. Environmental Protection Agency). 2002a. A Review of the Reference Dose and Reference Concentration Process. Final Report. EPA/630/P-02/002F. Risk Assessment Forum, U.S. Environmental Protection Agency. December, 2002.

EPA (U.S. Environmental Protection Agency). 2002b. Perchlorate Environmental Contamination: Toxicological Review and Risk Characterization. External Review Draft. NCEA-1-0503. National Center for Environmental Assessment, Office of Research And Development, U.S. Environmental Protection Agency, Washington, DC. [Online]. Available: http://cfpub.epa.gov/ncea/cfm/recordisplay.cfm?deid=24002 [accessed October 6, 2004].

EPA (U.S. Environmental Protection Agency). 2002c. Report on the Peer Review of the U.S. Environmental Protection Agency's Draft External Review Document "Perchlorate Environmental Contamination: Toxicological Review and Risk Characterization". EPA/635/R02/003. June, 2002. National Cencter for Environmental Assessment, Office of Research and Development, U.S. Environmental Protection Agency, Washington, DC. [Online]. Available: http://www.epa.gov/ncea/pdfs/perchlorate/final_rpt.pdf [accessed August 23, 2004].

Fisher, J., P. Todd, D. Mattie, D. Godfrey, L. Narayanan, and K. Yu. 2000. Preliminary development of a physiological model for perchlorate in the adult male rat: a framework for further studies. Drug Chem. Toxicol. 23(1):243-258.

Fisher, J.W., T.A. Whittaker, D.H. Taylor, H.J. Clewell, III, and M.E. Andersen. 1989. Physiologically-based pharmacokinetic modeling of the pregnant rat: A multiroute exposure model for trichloroethylene and its metabolite, trichloroacetic acid. Toxicol. Appl. Pharmacol. 99(3):395-414.

Fisher, J.W., T.A. Whittaker, D.H. Taylor, H.J. Clewell, III, and M.E. Andersen. 1990. Physiologically-based pharmacokinetic modeling of the lactating rat and nursing pup: A multiroute exposure model for trichloroethylene and its metabolite, trichloroacetic acid. Toxicol. Appl. Pharmacol. 102(3): 497-513.

Greer, M.A., G. Goodman, R.C. Pleus, and S.E. Greer. 2000. Does environmental perchlorate exposures alter human thyroid function? Determination of the dose-response for inhibition of radioiodine uptake. [Abstract]. Endocr. J. 40(Supp 1):146.

Greer, M.A., G. Goodman, R.C. Pleus, and S.E. Greer. 2002. Health effects assessment for environmental perchlorate contamination: The dose response for inhibition of thyroidal radioiodine uptake in humans. Environ. Health Perspect. 110(9):927-937.

Hays, M.T. and D.H. Solomon. 1965. Influence of the gastrointestinal iodide cycle on the early distribution of radioactive iodide in man. J. Clin. Invest. 44: 117-127.

Hollowell, J.G., N.W. Staehling, W.H. Hannon, D.W. Flanders, E.W. Gunter, G.F. Maberly, L.E. Braverman, S. Pino, D.T. Miller, P.L. Garbe, D.M. DeLozier, and R.J. Jackson. 1998. Iodine nutrition in the United States. Trends and public health implications: Iodine excretion data from National Health and Nutrition Examination Surveys I and III (1971-1974 and 1988-1994). J. Clin. Endocrinol. Metab. 83(10):3401-3408.

Kamm, G., and G. Drescher. 1973. Demonstration of perchlorate in urine [in German]. Beitr. Gerichtl. Med. 30:206-210.

Merrill, E.A. 2000. Consultative letter, AFRL-HE-WP-CL-2000-0036. Human PBPK Model for Perchlorate Inhibition of Iodide Uptake in the Thyroid. Memorandum with attachments to Annie M. Jarabek, NCEA, U.S. Environmental Protection Agency, Research Triangle Park, NC, from Elaine Merrill, Air Force Research Laboratory/HEST, Department of the Air Force, Wright-Patterson Air Force Base, OH. June 28, 2000.

Merrill, E.A. 2001a. Consultative letter, AFRL-HE-WP-CL-2001-0005. PBPK Model for Iodide Kinetics and Perchlorate-Induced Inhibition in the Male Rat. Memorandum with attachments to Annie M. Jarabek, NCEA, U.S. Environmental Protection Agency, Research Triangle Park, NC, from Elaine Merrill, Air Force Research Laboratory/HEST, Department of the Air Force, Wright-Patterson Air Force Base, OH. May 8, 2001.

Merrill, E.A. 2001b. Consultative letter, AFRL-HE-WP-CL-2001-0008. PBPK Model for Perchlorate-Induced Inhibition of Radioiodide Uptake in Humans. Memorandum with attachments to Annie M. Jarabek, NCEA, U.S. Environmental Protection Agency, Research Triangle Park, NC, from Elaine Merrill, Air Force Research Laboratory/HEST, Department of the Air Force, Wright-Patterson Air Force Base, OH. June 5, 2001.

Merrill, E.A., R.A. Clewell, J.M. Gearhart, P.J. Robinson, T.R. Sterner, K.O. Yu, D.R. Mattie, and J.W. Fisher. 2003. PBPK predictions of perchlorate distribution and its effect on thyroid uptake of radioiodide in the male rat. Toxicol. Sci. 73(2):256-269.

O'Flaherty, E. J., W. Scott, C. Schreiner, and R. P. Beliles. 1992. A physiologically-based kinetic model of rat and mouse gestation: disposition of a weak acid. Toxicol. Appl. Pharmacol. 112(2):245-256.

Reidel, C., O. Dohan, A. De la Vieja, C.S. Ginter, and N. Carrasco. 2001a. Journey of the iodide transporter NIS: From its molecular identification to its clinical role in cancer. Trends Biochem. Sci. 26(8):490-496.

Reidel, C., O. Levy, and N. Carrasco. 2001b. Post-translational regulation of the sodium/iodide symporter by thyrotropin. J. Biol. Chem. 276(24):21458-21463.

Shelley, M.L., M.E. Andersen, and J.W. Fisher. 1988. An inhalation distribution model for the lactating mother and nursing child. Toxicol. Lett. 43(1-3):23-29.

Soldin, O.P., S.J. Soldin, and J.C. Pezzullo. 2003. Urinary iodine percentile ranges in the United States. Clin. Chim. Acta 328(1-2):185-190.

Stanbury, J.B., and J.B. Wyngaarden. 1952. Effect of perchlorate on the human thyroid gland. Metabolism 1(6):533-539.

Williams, L.R., and R.W. Leggett. 1989. Reference values for resting blood flow to organs of man. Clin. Phys. Meas. 10(3):187-217 (as cited in Brown et al. 1997).

Wolff, J. 1998. Perchlorate and the thyroid gland. Pharmacol. Rev. 50(1):89-105.

Yu, K.O., D.A. Mahle, L. Narayanan, R.J. Godfrey, P.N. Todd, P. Parish, J. MacCafferty, T. Ligman, T. Sterner, G. Buttler, P.N. Todd, M.A. Parish, J.D. McCafferty, T.A. Ligman, C.D. Goodyear, T.R. Sterner, T.A. Bausman, D.R. Mattie, and J.W. Fisher. 2001. Tissue distribution and inhibition of iodide uptake by perchlorate in pregnant and lactating rats in drinking water studies. Toxicologist 60(1):291-292.

Yu, K.O., L. Narayanan, D.R. Mattie, R.J. Godfrey, P.N. Todd, T.R. Sterner, D.A. Mahle, M.H. Lumpkin, and J.W. Fisher. 2002. The pharmacokinetics of perchlorate and its effect on the hypothalamus-pituitary-thyroid axis in the male rat. Toxicol. Appl. Pharmacol. 182(2):148-159.